Uncommon Sense

Uncommon Sense

The Heretical Nature of Science

Alan Cromer

OXFORD UNIVERSITY PRESS

New York   Oxford

Oxford University Press

Oxford   New York
Athens   Auckland   Bangkok   Bombay
Calcutta   Cape Town   Dar es Salaam   Delhi
Florence   Hong Kong   Istanbul   Karachi
Kuala Lumpur   Madras   Madrid   Melbourne
Mexico City   Nairobi   Paris   Singapore
Taipei   Tokyo   Toronto

and associated companies in
Berlin   Ibadan

Library of Congress Cataloging-in-Publication Data
Cromer, Alan. 1935–
Uncommmon sense / the heretical nature of science / Alan Cromer.
p.   cm.   Includes bibliographical references and index.
ISBN 0-19-508213-3
ISBN 0-19-509636-3 (Pbk.)
1. Science—Philosophy.   2. Science—History.   3. Thought and thinking.
I. Title.
Q175.C897   1993
501—dc20   93-22892

10 9 8 7 6 5 4 3 2

Printed in the United States of America

*To my sister Barbara,
my brother Richard,
and my wife, Janet*

Preface

It has been less than 400 years since Galileo's telescope and Newton's mathematics opened the heavens to human investigation, little more than 100 years since the realization that human beings are the result of an evolutionary process that has extended over unthinkable millions of years, and less than 40 years since the discovery of the molecular structure of the genes that are the basis of all life. In contrast with the relative ignorance in which human beings have lived for most of their 200,000 years on earth, ours is an an age of spectacular knowledge and insight.

There are essentially two views of how we acquired this knowledge. The traditional view is that it was a product of the special genius of ancient Greece, that Greece, alone of all the civilizations on earth, developed the concepts of objectivity and deductive reasoning that are necessary for science. The modern view, which took shape in this century as scholars learned more about the mathematics of ancient Babylonia and the science of medieval China, is that science develops in every civilization. It may be more brilliant in one civilization than another, and perhaps it was especially brilliant in Greece, but science of some sort has been part and parcel of the normal development of humankind.

The difficulty with the modern view is that it can't explain why, having evolved 200,000 years ago, human beings achieved real scientific knowledge only in the last few hundred years. It's as though there is some great barrier to human thought that has only recently been penetrated.

From the work of the Swiss psychologist Jean Piaget, we know that human beings have a fundamentally egocentric conception of the world. Growing up in modern society means learning to accept the existence of an external world separate from oneself. It is hard. Most of humankind, for most of its history, never learned to distinguish the internal world of thoughts and feelings from the external world of objects and events. Each individual was indivisibly connected to the natural and supernatural powers that govern the world. Cutting this connection, which is necessary before science can develop, goes against the grain of human nature.

This brings us back to the traditional view, at least in the sense that science is very special—that it isn't part and parcel of human development. But then why did human beings develop a large brain, if it wasn't to cope better with the external world? And having developed a large egocentric brain, how was the external world ever discovered? These are some of the questions addressed in this book.

Our working hypothesis is that the antecedents of science either permeate a culture or are absent altogether. This follows from the Piagetian view that concepts build in stages. A culture that eventually developed science must have included, in its prescientific era, a number of activities that taught the mind to differentiate between internal thought and external reality. A culture that never developed science won't contain such activities; indeed, all its activities will perpetuate the confusion of thought and reality. This hypothesis is supported by an examination of the cultures of ancient Greece and Israel.

Already in the *Iliad* we find the tradition of debate teaching the Greeks the art of persuasion, and in the Old Testament we find the tradition of prophecy perpetuating the Jewish penchant for piety. Debate requires the existence of independent minds, just as prophecy requires the existence of men who communicate directly with God. Each is based on a fundamentally different notion of how human beings acquire knowledge. With debate we have a major step toward the separation of the internal and external worlds that is essential to science.

From the sixth century B.C. to the second century A.D., Greek mathematics and astronomy rose to a height that wasn't to be duplicated for another 1,400 years. Little of this work was done in Greece proper. After the army of Alexander the Great brought Greek culture to North Africa and Asia in the fourth century B.C., much of Greek mathematics and astronomy were carried out in Alexandria.

For 500 years after Alexander, Greek culture dominated Europe and western Asia. Then, with the rise of Christianity, egocentrism replaced objectivity throughout Europe. In western Asia, Islam maintained interest in Greek science for many centuries, until it, too, succumbed to the allures of the supernatural. But by then, Greek science had been reintroduced to Europe.

According to the neotraditionalist interpretation of science presented in this book, not only is Greece restored as the undisputed fountainhead of science, but also no other civilization is considered to have made any attempt to separate internal thought from external reality. Although China and India played essential roles in the subsequent development of science, neither developed objectivity.

The unique origin of science in Greek culture doesn't make science any less ecumenical, since that culture was multinational. Mathematicians, astronomers, and scholars from many countries worked and studied in the library and museum in Alexandria and in other centers of learning throughout the Hellenistic world. And after the rise of Christianity in the West destroyed the last traces of Greek learning in Europe, the flame was kept alive in India and Arabia later to ignite the European Renaissance.

The notion that science and objective thinking are unnatural human activities seems quite radical at first. But when you think about it, monogamy, honesty, and democratic government are unnatural human behaviors as well. We are truly a species that has invented itself out of rather unpromising material. Our only claim to greatness is that we have at times gone against the grain of our own egocentrism to forge a higher vision of the world.

A Look Ahead

The current status of science is reviewed in Chapter 1. In spite of some of its early mistakes, science is found to be overwhelmingly cumulative rather revolutionary, building a body of permanent knowledge, most of it very recent. The very recentness of this knowledge leads us in Chapter 2 to an investigation of human thought, where the main conclusion is that human beings are inherently egocentric in their thinking, with little inclination toward belief in an external world independent of their wishes. Yet we are able to "do" science today. What is the origin of this capacity? What were the environmental pressures that channeled evolution in our direction? Why does a cave dweller need the mind of an Einstein? These are deep questions, at the edge of our understanding. Chapter 3 reviews the current state of our knowledge of human evolution for possible answers. We will find that whatever evolution was after, it certainly wasn't science. It was, however, something that can be put into the service of science under the right circumstances.

What were these circumstances? How rare are they in human history? Chapter 4 traces the roots of Greek rationality in the oldest Greek literature, the *Iliad* and the *Odyssey*. Here, among Homer's stories of gods and battles, are some of the key concepts necessary for science and mathematics, such as open debate and noncontradiction. By contrast, the prophetic traditions of the Old and New Testaments represent the egocentric view that knowledge is belief. Of course, many factors besides Homer were necessary for the development of science, including the survival of Greek learning for 1,000 years.

Chapter 5 presents a brief review of the history of science from ancient Babylonia to Copernicus. One purpose of this chapter is to show just how different scientific thought is from ordinary thought. Even the earliest theorems of Greek mathematics are subtle and elusive. Another purpose is to trace the spread of Greek culture into North Africa, western Asia, and India. It is a central thesis of this book that science has a unique historical origin in ancient Greece and that all other civilizations, down to our own, learned science directly from Greek science. There is much controversy on this point and many claims to the contrary. Can ancient Greece really have been as unusual as all that? What about the great civilizations of China and India? What about Arab science?

The Arabs originally learned about Greek science from the Indians and then directly from the works of Euclid and Ptolemy. Chapter 6 examines the civilizations of traditional China and India for traces of original science. Indian science is shown to be a bowdlerized version of Greek science; it had the substance but not the form. China produced many major technologies vital to modern science—principally paper and printing—but not science itself. This failure demonstrates that science can develop only under a very precise set of cultural circumstances.

Only with the scientific revival in Europe in the sixteenth and seventeenth centuries did science push through the limits of Greek science. Why? What was so special about Europe? This is a critical question, especially during a period when it's fashionable to favor every culture but our own. It is explored in Chapter 7, which examines the historical and cultural roots of modern science. Again, we find that a combination of rather unlikely factors is required to explain the burst of scientific activity during and after the "scientific revolution." Prominent among these factors was the number of people involved. You can't do science without scientists.

This brings us to the question, avoided up to now, of what science actually is. How does it differ from other intellectual activities? Is it its subject matter or its methods? In Chapter 8 we adopt John Ziman's social definition of science and show how it fits into the cultural and historical interpretation of this book. Scientists are usually suspicious of sociological analyses of their work, since some sociologists have used them to devalue science. But that isn't the case here. On the contrary, the sociological definition of science gives us a content-free way to distinguish science from pseudoscience, as a number of case studies show.

The success of science in learning about the external world is our best proof that there is an external world. But if there is, then it is what it is. This is such a recent idea that its implications are not generally accepted. For example, the claim that space is the next frontier for humankind is based on a belief in unlimited progress, a notion contrary to that of an external reality. Chapter 9 explores several questions related to the limits of progress, including whether we are alone in the universe and whether we will ever be able to receive communication from life on another planet.

The recentness of science also has implications for how it is taught. Science isn't about the meaning of words, but about things in themselves. This must be learned, in the first instance, by direct experience. Because teachers weren't taught in this way, they don't teach this way to their students. Chapter 10 examines what is being done to enhance the skills of existing science teachers and to improve the education of future teachers. It also tackles the problem of our deteriorating public-school system and suggests what can be done so that it can provide a quality education for all.

A Personal Note

A word about my background may help the reader sort out my prejudices. I am a theoretical nuclear physicist who has taught graduate and undergraduate physics for over thirty years. I have been in love with science ever since my pediatrician first carried me into the laboratory next to his examining room. Currently I am engaged in projects to enhance the knowledge of middle-school science teachers and to improve the teaching of science to middle-school students.

As a child, I attended a reformed Jewish synagogue founded by German Jews in Chicago. In the nineteenth century, the Jews of Germany had experienced a great renaissance. They acquired German citizenship, won admittance to German universities, and even obtained professorships. Somehow I absorbed more of this Hellenized Judaism than I ever did of traditional Judaism. When I was told the story of Joshua destroying the walls of Jericho, I identified with the defending Canaanites, not with the attacking Israelite tribesmen. It was at the same time that the Jews of Europe were being destroyed by the atavistic tribalism of Germany.

I believe that rational civilization, with its science, arts, and human rights, is humankind's greatest hope for nobility. But like Jericho, it's but an oasis in the midst of a vast desert of human confusion and irrationality.

Boston   A. C.
December 1992

Acknowledgments

I first studied science with my sister, Barbara, who taught me chemistry as we did the dishes together. Although my brother, Dick, could never be found at that time, more recently he has given me invaluable support and advice for this book. He made me get rid of fancy dictionary words in favor of a more direct and readable style.

Parts of the manuscript were read by Michael Sobel, Bernard Gottschalk, Eberhard von Goeler, and Janet Cromer. Their insights and corrections make this a more accurate and more readable work.

I am grateful to my colleagues Michael Silevitch and Christos Zahopoulos, whose tireless efforts on behalf of science education have been both inspirational and informative, and to the middle-school science teachers in our summer programs, whose knowledge and experience of the educational process I've freely drawn on in this book. My undergraduate students, the source of much joy and frustration, provided the raw material on which the basic premise of this book is based.

But nothing would have been possible without my wife, Janet, whose love and support provided the safe haven in which I could indulge myself with unusual ideas and speculations. She also taught me that there are times to stop speculating and come to bed.

Contents

Uncommon Sense

1
Aspects of Science

Ages hence will remember our century not for its barbaric wars, but for its major contributions to fundamental scientific knowledge. In this century, genetics has been reduced to chemistry, and chemistry to the quantum physics of atoms and molecules. Even the processes that created the atoms themselves in the interior of ancient stars are well understood. Whatever the scientific discoveries of future ages, they will be based on the vast body of twentieth-century science.

Yet for all its intellectual and practical achievements, science is poorly understood by lay people and scientists alike. Most people believe, for instance, that science—whatever it is—arose from humankind's innate intelligence and curiosity, that it is a natural part of human development and, as such, probably began in prehistoric times "whenever and wherever men tried to solve the innumerable problems of life" (Sarton, 1960, p. 3), that from the day a human ancestor first broke a rock to cut a hide, it became inevitable that someday his descendants would land on the moon and that new scientific knowledge will continue to replace old knowledge as science develops in the future.

But my reading of history and of theories of human development points to very different conclusions. As this book will argue, scientific thinking, which is analytic and objective, goes against the grain of traditional human thinking, which is associative and subjective. Far from being a natural part of human development, science arose from unique historical factors. And viewed against the thousands of years of human existence, science is very recent.

The Recentness of Science

Scarcely 200 years have passed since that November day in 1783 when Pilátre de Rozier and the marquis d'Arlandes became the first human beings to fly, soaring aloft in a Montgolfier brothers' hot-air balloon before the 200,000 amazed spectators who filled the gardens of the Tuileries. Among the spectators was the novelist Sebastien Mercier, who reported on the event:

> What with the novelty, the dignity of the experiment, the unclouded sky, welcoming as it were the travellers to his own element, the attitude of the two men sailing into the blue, while below their fellow-citizens prayed and feared for their safety, and lastly the balloon itself, superb in the sunlight, whirling aloft like a planet or the chariot of some weather-god—it was a moment which never can be repeated, the most astounding achievement the science of physics has yet given to the world. (Quoted in Clark, 1979, p. 30)

The advent of human flight instantly became part of the European imagination. Romantic novels and plays based on the adventures of aeronauts appeared in abundance by 1784, and in 1786 the first space opera, *Die Luftschiffer,* was performed in St. Petersburg. Philosophers, poets, and statesmen all had something to say about this remarkable event. Thomas Jefferson suggested that balloons could be used to travel to the North Pole, and Diderot believed that balloon flight would one day enable human beings to travel to the moon.

According to the literary historian I. F. Clarke, the very concept of an ever-improving future came into existence with balloon flight:

> Ever since the first balloon ascents of 1783 and the first wave of balloon stories the course of futuristic fiction has been the record of an ever closer and ever more complicated relationship between science and society. Nowadays, when our frequent anxieties about the consequences of technological inventiveness cause many to fear the future, it is difficult to enter into the far freer and far more hopeful spirit of the time when the new sciences seemed to promise the most desirable improvements in the condition of mankind. (1979, p. 33)

In many ways, the decade of the 1780s divides the past from the present. In science and technology, it saw the beginning of human flight, the Watt steam engine, the birth of modern chemistry, and Jenner's small pox vaccine. In politics, it saw the French Revolution and the writing of the United States Constitution. In fashion, the 1790s saw men abandon their aristocratic accouterments of wig, sword, and embroidered coat for a simpler country look.

From this period on, human culture has been driven by the forces of science and invention, which grow stronger with each new discovery. Our current technology and way of life differ more from that of the eighteenth century than the eighteenth century's did from that of the time of the pharaohs. Although this is generally recognized, the very recentness of our vast increase in scientific knowledge distorts our perspective on its permanent significance.

How durable is the scientific knowledge acquired in the last 300 years? Will scientists 2,000 years from now still use Newton's laws for mechanics and the periodic table for chemistry? The history of science contains a number of instances when new theories replaced old ones. The scientific revolution itself began with the sun-centered planetary system of Copernicus, which replaced the 2,000-year-old earth-centered system of Hipparchus and Ptolemy, and it was completed by the mechanics of Newton, which replaced the equally ancient physics of Aristotle. This period, which spanned about 150 years, ended 300 years ago. Subsequently, Lavoisier replaced the Aristotelian elements (air, water, earth, and fire) with chemical elements, and Darwin replaced divinely created species with the products of natural selection. Is all of science just an unending cycle of death and rebirth? Will today's theories strut their hour upon the stage and then be heard no more?

There is a respectable body of opinion, stemming from the work of Thomas Kuhn, that "portrays scientific development as a succession of tradition-bound periods punctuated by non-cumulative breaks" (1970, p. 208). During these breaks, or revolutions as Kuhn calls them, a new system of ideas replaces an old system by a process outside the scope of the old system. The best example of this is the replacement of the whole system of ideas associated with Aristotle by that associated with Newton. Aristotelian physics is the best, and perhaps the only, example of a closed, tradition-bound system of ideas that is inherently incapable of modifying itself. That's why we call the scientific revolution a revolution.

The development of relativity and quantum mechanics in the

first quarter of this century amazed the world of physics with new concepts and principles unimaginable in Newtonian mechanics. Does this make Newtonian mechanics just another tradition-bound system, more powerful than its predecessor but destined to be replaced by still more powerful systems in the future? This is the way it may have looked when these theories were new, but with the benefit of a little more historical distance it looks quite different.

First, relativity and quantum mechanics have stood the test of time, in spite of diligent efforts to find alternative formulations. Second, relativity and quantum mechanics never replaced Newtonian mechanics. Physics students study Newtonian mechanics as thoroughly as ever and, in addition, relativity and quantum mechanics. Without replacement, the developments of relativity and quantum mechanics don't meet Kuhn's own definition of revolutions as "noncumulative breaks." Removing them from the list changes our whole perspective on science because we are then without a bona fide revolution for the last 150 years.

Although 150 years is short in terms of human history, it is a respectable fraction of the entire age of modern science. Seen from this perspective, we can change our metaphor from political upheavals to biological growth. The mistakes of the early days of science are, like the falls and bumps of a toddler, a nonrecurring phase of the growing-up process.

Science is overwhelmingly cumulative, not revolutionary, in its structure. This means that most of its established results—even those established recently—will be around forever. A particular result may be found to be an instance of a more general result, but its factualness, as far as it goes, will never change. Archimedes' principle of buoyancy is still taught today in much the same way as it was first stated 2,300 years ago, although we now derive it from more general principles. Indeed, most of Greek science and all of its mathematics are still as valid as ever. It is the physics of Aristotle that is anomalous.

The Completeness of Science

Once we get over the distortion of perspective caused by being so close in time to so much new scientific knowledge, we can see that much of it is already essentially complete. By this I mean that the experimental and theoretical basis of some of our fundamental knowledge is so extensive that there is little likelihood of its being changed to any significant degree. This is an astonishing assertion, given the breathtaking pace of discovery today. But the pace of discovery is possible precisely because our fundamental knowledge is so complete.

Take, for example, our knowledge of the chemical elements. It was only in 1783, a few months before the ascent of de Rozier and d'Arlandes, that Antoine-Laurent Lavoisier announced to the Academy of Sciences in Paris that water is a compound composed of hydrogen and oxygen.[1] Until then, chemists had clung tenaciously to the ancient belief that all matter was composed of a combination of the Aristotelian elements. Lavoisier proposed instead that an element be defined experimentally as any substance that can't be decomposed by chemical methods.

John Dalton, an English schoolteacher, suggested in 1803 that perhaps each element is composed of its own kind of atoms. In a chemical reaction, the atoms of different elements combine to form molecules, the smallest unit of a compound. For example, an atom of hydrogen might combine with an atom of oxygen to form a molecule of water. Experimentally it is found that one gram of hydrogen always combines with eight grams of oxygen to form nine grams of water. Given this weight ratio of 8:1 and assuming that each molecule of water consists of one oxygen atom and one hydrogen atom, an oxygen atom must weigh eight times as much as a hydrogen atom. In fact, a molecule of water consists of one oxygen atom and two hydrogen atoms, so that each oxygen atom is sixteen times as heavy as each hydrogen atom.

By measuring the weight ratios of the elements in a wide variety of compounds, chemists gradually established a system of atomic weights. In this system, an atom of carbon is assigned the weight of 12. An oxygen atom, being found to be one-third heavier than a carbon atom, is assigned a weight of 16. Hydrogen, being one-sixteenth the weight of oxygen, is assigned the weight of 1. Determining the weights of the elements was one of the major tasks of quantitative chemistry in the nineteenth century.

Twenty-three elements were known in Lavoisier's time, but by 1869, nearly sixty elements and their weights were known. It was found that the properties of the elements don't change steadily with weight, but that certain properties recur periodically. For example, the highly reactive elements lithium, sodium, and potassium have weights (7, 23, and 39, respectively) that differ by the same amount. (That is, the weight difference 23 minus 7 between sodium and lithium is the same as the weight difference 39 minus 23 between potassium and sodium.) In 1869, Dmitry Mendeleyev succeeded in arranging by weight the sixty known elements in a table with fixed rows and columns so that elements with similar properties were in the same column. He achieved this by boldly allowing gaps in the table for yet undiscovered elements. As these missing elements were discovered, they were found to have the weights and properties predicted by their place in the table.

The periodic table is a great achievement of experimental chemistry. That seven elements occur between lithium and sodium, and seven between sodium and potassium, is a fact. However, without understanding why these similar elements are separated by just seven elements, we can't rule out the possibility of someday discovering an eighth element between them.

Such understanding came during the period 1911 to 1938. Experimental studies showed that an atom's mass is almost all concentrated in a tiny positively charged region called the nucleus, which is orbited by much lighter negatively charged electrons. The number of electrons determines the chemical element: Hydrogen has one electron; helium, two; lithium, three; sodium, eleven; and so on. The motion of these electrons was found not to be correctly described by Newtonian mechanics. A new theory of motion, called quantum mechanics, applicable to atomic-sized particles, was developed in 1925. This theory describes precisely how the electrons in atoms are configured into concentric shells and why, for example, the atoms with two, ten, eighteen, thirty-six, fifty-four, and eighty-six electrons are inert gases (helium, neon, argon, krypton, xenon, and radon) and the atoms with one, three, eleven, nineteen, thirty-seven, fifty-five, and eighty-seven electrons are the very reactive alkali halides (hydrogen, lithium, sodium, potassium, rubidium, cesium, and francium). Quantum mechanics also allows physicists to calculate the wavelengths of the light emitted when an electrical current passes through the vapors of an element (as in a "neon" sign). The agreement between the measurements of thousands of such wavelengths and the theoretical calculations confirmed the validity of the theory.

The elements in the periodic table are actually in the order of the number of electrons they carry—their atomic number. This order is virtually the same as their ordering by weight. But the weight comes almost entirely from the nucleus. In the 1930s it was discovered that the atomic nucleus is composed of positively charged protons—equal to the number of electrons—and electrically neutral neutrons. Each proton and neutron weighs about 1 (on the scale in which a carbon atom weighs 12), whereas an electron weighs only 0.0005. The nucleus of a hydrogen atom is a single proton, so hydrogen weighs 1 (approximately). The nucleus of a carbon atom contains six protons and six neutrons, so a carbon atom weighs 12; the nucleus of an oxygen atom contains eight protons and eight neutrons, so a carbon atom weighs 16. Thus the weights of atoms, as originally inferred from chemical analysis, are now understood in terms of the structure of the atomic nucleus.

In the late 1930s it became clear that nuclei existed with the same number of protons but different numbers of neutrons. Such nuclei are called isotopes. For example, a hydrogen nucleus can be a single proton, a proton bound to one neutron, or a proton bound to two neutrons. All three are orbited by a single electron and are chemically very similar. However, the atom with one proton and one neutron in its nucleus weighs 2, and the atom with one proton and two neutrons weighs 3. Usually a particular isotope is specified by giving the name of the element followed by the total number of protons and neutrons—for example, hydrogen-1, hydrogen-2, hydrogen-3.[2]

Every element is now known to have a number of isotopes, though usually one or two are much more common than the others. For example, only 1 in 50,000 hydrogen atoms has a nucleus with a neutron in it. The common isotope of carbon (carbon-12) has six protons and six neutrons, and 1 in 100 carbon atoms has a nucleus with six protons and seven neutrons (carbon-13). Carbon-14, which has six protons and eight neutrons, is radioactive and present in only trace amounts.

There are over 100 elements, and they can be arranged according to the number of protons in their nucleus. The first ninety-two elements—from hydrogen (with one proton) to uranium (with ninety-two protons)—occur naturally on earth. However, four of these—technetium, promethium, astatine, and francium—are so rare that weighable quantities of them must be produced synthetically.[3] All the elements with more protons than uranium has, such as neptunium (ninety-three protons) and plutonium (ninety-four protons), are also produced synthetically.

Synthetic elements up to curium (ninety-six protons) can be made in gram and kilogram amounts in nuclear reactors, and microgram amounts of berkelium and californium (ninety-seven and ninety-eight protons) have been obtained after years of work. Only a few atoms—and they are radioactive with very short lives—have been produced with ninety-nine to 109 protons (Armbruster and Münzenberg, 1989).[4] Individual atoms with more than 109 protons may some day be created in a heavy-ion accelerator, but they will survive for only a few thousandths of a second. So in the ordinary sense of the term, the elements end with californium (ninety-eight protons).

We now can say that our knowledge of the elements is essentially complete in the sense that all the elements from hydrogen to californium have been isolated and studied and we know that we haven't missed any. This last comes from our understanding of the structure of atoms in terms of electrons, protons, and neutrons. There are exactly seven elements between lithium and sodium, because lithium has three electrons and sodium has eleven, and there are exactly seven integers between 3 and 11 (4, 5, 6, 7, 8, 9, and 10). Our knowledge of the ninety-eight elements is as complete as our knowledge of the numbers from 1 to 98.

This review of atomic theory is presented as a concrete example of an important body of scientific knowledge that is both recent and essentially complete. No scientist questions the validity of this knowledge—indeed, few could do their work without it. For example, isotopes, although they were discovered just sixty years ago, are essential to many areas of scientific research, from geology to molecular biology.

This fundamental knowledge about the structure of matter gives scientists a solid base on which to advance their understanding of the structure of stars, the creation of the elements, and the evolution of life. At the same time, such knowledge limits the possible. This is a difficult concept to grasp, because the stupendous achievements of our technology engender the belief that anything is possible. So a president can believe in the possibility of shielding a continent from attack by intercontinental missiles, and a UFOlogist can believe in the possibility of interstellar travel. But just as there are no undiscovered continents on earth, so there are no undiscovered elements with fewer than ninety-eight protons. Everything on earth is made of the same ninety-eight elements that naturally exist or can be artificially produced in weighable quantities.

But as important as the elements are to our understanding of nature, we must be careful not to confuse the chemical elements as we know them with the ultimate elements of nature. Almost by definition, nothing can be known about the ultimate elements, whatever that term means. What Lavoisier did was to stop speculating about ultimate elements and start thinking about chemical elements—substances in his laboratory that couldn't be refined into anything simpler. It is these substances that turned out to have ordinality—that is, could be ordered 1, 2, 3, . . . , so that we could know we weren't skipping any.

We still know little or nothing about the ultimate elements. It is tempting to say that the electrons, protons, and neutrons that compose all atoms are ultimate, but neutrons and protons themselves have structure. As physicists have penetrated this structure, a plethora of mesons, baryons, leptons, quarks, and gluons have emerged, rivaling the number of chemical elements. The current "standard model" uses forty-five or so "fundamental particles" to explain the proton, the neutron, and several hundred other subnuclear particles created in high-energy accelerators. That such a menagerie constitutes the ultimate elements of nature is most unsatisfying, but it's the best we can do at present. It may be that we can never have complete knowledge of the Ultimate. But we can have complete knowledge of important aspects of the physical world. The properties of electrons, protons, and neutrons—the basic constituents of atoms—are known in great detail. No future theory will replace this knowledge; indeed, this knowledge is what any deeper theory will be called upon to explain.

We already have the example of relativity and quantum mechanics incorporating Newtonian mechanics into theories of much greater scope and validity. Newtonian mechanics remains as valid in its domain as plane geometry is in its. A surveyor is perfectly justified in using the theorems of plane geometry to lay out the foundation of a building, even though the earth is a sphere, because over a small area the surface of a sphere approximates a plane. Over a large area, of course, plane geometry is not satisfactory, and a system of latitudes and longitudes must be adopted. This system, which is the correct geometry of a sphere, includes plane geometry as a special case when the area involved is small.

In much the same sense, Newton's laws of motion and of gravity, though a special case of the more general theories of quantum mechanics and relativity, constitute a complete and consistent theory. Although not valid for the motion of electrons inside an atom or for the gravitational field near a black hole, Newtonian mechanics does apply to everything in between, from the motion of a Ping-Pong ball to that of a galaxy. This is why speculations that contradict it, such as the idea of Velikovsky (1950) that the earth actually stopped spinning when Joshua made the sun stand still, can be given no scientific credence.

Newtonian mechanics replaced Aristotelian metaphysics with a mathematical theory of functional relationships. This theory contains certain fundamental relations, such as the relation between the motion of an object and the force acting on it and the relation between this force and the object's position, from which the position of the object at any time (the functional relationship between position and time) can be calculated. Our belief in the validity of the scheme comes from its simplicity, its inner consistency, and the exact correspondence between calculated and observed motions.

The fundamental relationships of the theory are the so-called laws of physics, a term that means no more or less than that the relations have been found to have universal applicability over a large range of conditions. That Newtonian mechanics is in fact a special case of the still more general laws of relativity and quantum mechanics doesn't invalidate it, any more than spherical geometry invalidates plane geometry. Rather, Newtonian mechanics is enhanced by being firmly embedded in a grand theoretical scheme that correctly describes everything from the motion of protons inside a nucleus to the expansion of the universe itself. Likewise, any future development that shows that this grand scheme is itself a special case of a supergrand theory will enhance the credibility of the grand theory within its range of validity. Our level of understanding of the laws of physics is such that newer theories may enfold older ones, but they won't destroy them. It is in this sense that Newtonian mechanics, quantum mechanics, and relativity are said to be essentially complete.

This doesn't preclude the possibility of future theoretical development in these fields, but such development will be about things that we know we don't know, not reversals of things that we do know. For example, Newtonian mechanics contains one universal constant, G, that determines the magnitude of the gravitational force between two masses. As far as we know, G is the same for all masses everywhere in the universe, now and in the past. It was first measured by Cavendish in the late eighteenth century, and all our knowledge of the masses of stars and planets, and of the weight that astronauts would have on the moon or on Mars, comes from our knowledge of G. It determines the astronomical scale of things. But we have no theory of why G is what it is.

Similarly, on the atomic scale we have a number of fundamental constants: the charge e of an electron, the mass m of an electron, the speed of light c, and the quantum constant h. Although all these can be measured, their values can't be—as yet—derived from any more fundamental theory. Still, from these constants the theories that we do have allow us to derive such things as the size of atoms, the binding of atoms into molecules, and the entire periodic table.

Thus to the question "Why are things the size they are?" we can answer "Because e, m, and h are the size they are." This isn't quite as silly as it sounds, for because everything is made of atoms, the sizes of everything are related. In particular, the distance between atoms in molecules and in solids is inversely related to the mass of an electron. If the electron were lighter, everything would be larger and less dense, and if the electron were heavier, everything would be smaller and denser. All chemistry would be affected by such changes, including the chemistry of life. For example, if the electron were 200 times its actual mass, two hydrogen-2 atoms would form such a small hydrogen molecule that their nuclei would fuse, thereby releasing nuclear energy. We know this because of experiments with muons, short-lived particles that behave like electrons with a mass 200 times the electron's mass (Rafelski and Jones, 1987).

Although some future theory may be able to relate G, m, e, c, and h to a smaller set of still more fundamental constants, it won't decrease the validity or scope of the existing relations that involve them. Our knowledge of Newtonian mechanics and quantum theory is complete because it is knowledge—not of absolute causes, but of relationships of broad generality—that is valid to the limits of our current measuring abilities (about one part in a million). Future theories can only broaden the scope and deepen the range of our knowledge.

Even such recent knowledge as the chemistry of genes is complete in some respects and is approaching completion in others. Genes, like atoms, were once elusive hypothetical constructs invented to explain a narrow range of phenomena. Today, the structures of many genes are known down to the last atom, and all the genes of all living organisms are known to be different orderings of the same four groups of atoms. Each gene is a segment of the complex deoxyribonucleic acid (DNA) molecule that resides inside every chromosome of every cell of every living organism.

In 1953, James Watson and Francis Crick determined that the DNA molecule consists of two long chains of atoms that spiral around each other—the famous double helix (Figure 1.1). The two chains

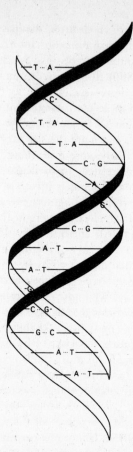

Figure 1.1 Section of a DNA molecule showing the two spiral structures connected by T–A and G–C base pairs. The sequence of these bases along one of these spirals codes the sequence of amino acids in a protein molecule. A base pair and the section of the spirals between pairs are composed of about fifty atoms of hydrogen, carbon, nitrogen, oxygen, and phosphorus.

are linked by groups of atoms, called bases, much like the steps of a spiral staircase. Each step consists of two bases that are attached tightly to their respective chain and weakly to each other. There are only four different bases in DNA, designated by the initials of their chemical names: A, T, G, C.[5] Furthermore, they pair up in a unique way: A always with T, and G always with C. Apart from this restriction, the pairs in a DNA molecule can be in any order. An endless variety of molecules are possible, differing only in the order of their linking pairs.

Figure 1.1 shows a section of a DNA molecule whose pairs, starting from the bottom, happen to be A–T, A–T, G–C, C–G, C–G, T–A, T–A, G–C, G–C, A–T, C–G, T–A, T–A, G–C, A–T, and so on. Consequently, along one chain the sequence of bases is AAGCCTTGGACTTGA . . . , and along the other strand there is the complementary sequence, TTCGGAACCTGAACT . . . There are two critical points here. First, every possible sequence of the letters A, T, C, G is a possible DNA molecule, so that the molecule looks like a coded message. Second, each half carries the same message, since A on one side always means T on the other. In this remarkable structure, we have the key to both reproduction and heredity. Hereditary information about producing an adult organism from a single cell is encoded in the ordering of the base pairs of the organism's DNA. Reproduction of a cell starts when the DNA's double chain unzips along the weak connection between its base pairs, forming two single chains. Each single chain then reconstitutes itself into a double chain when each base of the chain attracts its complementary base from the soup of unattached bases in which it dwells. Form and function are perfectly matched in the structure of DNA.

In human beings, the DNA in each cell is stored in forty-six chromosomes, twenty-three inherited from each parent. Each chromosome can be thought of as a book, with the base pairs being the letters in the book. Three consecutive bases along a chain form a word or a period marking the end of a sentence. Each sentence is a gene, and each gene is the code for the synthesis of a specific protein. Proteins are long strands of amino acids, and the sequence of the words in a gene determines the sequence of the amino acids in its protein. For example, the sequence beginning AAG CCT TGG is the code for a protein with the sequence of amino acids phenylaline–glycine–threonine, since AAG is the code for phenylaline, CCT is the code for glycine, and TGG is the code for threonine.

Proteins typically have about 100 amino acids, so each gene consists of about 300 base pairs. It is estimated that humans have about 100,000 genes. Thus some 30 million base pairs provide the code for making a human being out of amino acids. Human DNA contains about 3 billion base pairs, nearly 100 times more than seems to be needed. The genes, or functional segments of the DNA, are widely separated by regions with little known function. These nongene regions vary greatly from one individual to another and are the basis of the DNA identification techniques used in criminal investigations.

The set of 3 billion or so base pairs in human DNA constitutes the human genome. Although large, it is finite and modular. Efforts are under way to determine the base-pair sequences of the entire human genome. Depending on the funding, this project may take a decade or more. The more difficult task of recognizing all the genes may take longer. But in time we will know all the base-pair sequences of all the human genes, just as we now know all the chemical elements. This information won't fit on a wall chart, but it will be in a computer data base, accessible to researchers worldwide.

It will probably be less than 100 years from the Watson–Crick determination of the structure of DNA to the complete determination of the human genome. This is typical of the time it takes to complete a major area of science once the fundamental breakthrough has been made. Eighty-six years elapsed between Lavoisier's decomposition of water and Mendeleyev's periodic table, and fifty-six years between the periodic table and the quantum theory of atoms.

The Unity of Science

Our deep understanding of the molecular basis of heredity has removed any lingering doubts about the unity of science. In the nineteenth century, vitalists could believe that living organisms contained some vital essence that was different from the atoms of ordinary matter. But now we have an unbroken chain linking the physics of atoms to the structure of self-replicating molecules. No forces or essences not in the physicists' textbooks need be invoked to understand everything from the synthesis of the elements inside an exploding star to the evolution of life in ancient seas.

To be sure, the depth of this understanding isn't uniform. Nucleosynthesis, the creation of the heavier elements from primordial hydrogen and helium, is well understood. The measured abundances of the elements in the universe, as determined from astronomical observations, agree with detailed computer models of how these elements are formed in stars and dispersed back into space when stars explode. No similar models yet exist for the origin of life from the soup of complex molecules that rained on the oceans billions of years ago. Nevertheless, there is no doubt among scientists that when such a model is developed, it will be based on the known laws of physics and chemistry.

True Knowledge

These three aspects of science—its recentness, the completeness of some of its fundamental knowledge, and its intrinsic unity—mean that for the first time in human history we have true knowledge of the nature of existence and of our place in it. This fundamental fact is often ignored. Since science is necessarily tentative and uncertain at its growing edge, the great foundation of certainty that currently exists is seldom emphasized. Atoms and genes have changed from hypothetical notions into concrete objects whose existence is as certain as objects we can see and touch.

Academics cringe at the words *truth* and *certainty*. They believe that truth and certainty aren't possible because philosophers have shown that neither empirical nor deductive knowledge can be made error free. But in the case of a finite number of discrete entities, such as the chemical elements or the human genes, *certainty* is an appropriate word. And in any event, our knowledge of atoms and genes is as certain as our knowledge of tables and chairs, and a lot more certain than our knowledge of human behavior.

Academics periodically debate how much science should be required for a bachelor's degree in the humanities and social sciences. In arguing against the need for more than a token course in science, humanists stress the equivalent value of "poetic truth," and social scientists point out that science is just a white-male view of the world anyway. This mishmash of criticisms of science, in effect, devalues all knowledge. If the poets won't sing of science, who will?

In the past, science and technology were loudly hailed by the poets. In 1805 Wordsworth wrote:

> If the labours of men of science should ever create any material revolution, direct or indirect, in our condition, and in the impressions which we habitually receive, the Poet will sleep no more than at present; he will be ready to follow the steps of the Man of science, not only in those general indirect effects, but he will be at his side, carrying sensation into the midst of the objects of science itself. The remotest discoveries of the Chemist, the Botanist, or Mineralogist, will be as proper objects of the Poet's art as any upon which it can be employed. (Quoted in Clarke, 1979, p. 50)

How wrong he was! As science has pushed forward the frontiers of our understanding, poets, at least in the United States, have largely retreated into the inner world of their own psyches. Science has indeed completely changed the "condition, and . . . the impressions which we habitually receive," without producing any parallel changes in our arts or religions. Religion still rests its authority on the words of ancient prophets, and the arts, far from considering "the remotest discoveries of the Chemist . . . as proper objects of the Poet's art," don't even espouse such aesthetic ideals as harmony and restraint, suitable for an age of science.

Human Character

The reason for science's lack of approbation lies deep in the human character. Human beings, after all, love to believe in spirits and gods. Science, which asks them to see things as they are and not as they believe or feel them to be, undercuts a primary human passion. In every society, the skeptic is liable to be shouted down, if not stoned down, by the mob of believers. Our higher rational abilities don't develop spontaneously, but must be carefully cultivated by a process of formal education. And not just any formal education. Most of the educational systems developed over the millennia were designed to teach allegiance to the words of ancient texts, not a disinterested search for truth. That the truth lies in the book and not in nature is as old as the Vedic texts and as recent as Allan Bloom:

> Contrary to what is commonly thought, without the book even the idea of the order of the whole is lost. . . .
> I am not saying anything so trite as that life is fuller when people have myths to live by. I mean rather that a life based on the Book is closer to the truth, that it provides the material for deeper research in and access to the real nature of things. (1987, pp. 58, 60)

Science is the heretical belief that the truth about the real nature of things is to be found by studying the things themselves. Traditionally, scholarship has meant the study of books—the older, the better. The scientific triumphs of the last few centuries notwithstanding, this view still persists among many nonscientific academics. Few, however, are as blatant about it as Bloom is.

Views of Science

There is a profound paradox here. If human beings are basically tradition-bound, irrational creatures, how did science ever develop in the first place? The short answer is, "With difficulty." It wasn't the "natural" unfolding of the human potential, but the peculiar invention of a particular culture in a particular time. This view, once universally accepted, has been largely abandoned by scholars of this century. It has been replaced by either Kuhn's noncumulative revolutions or a bland relativism that applies the term *science* so indiscriminately as to hopelessly muddle thinking on the subject.

Historically, the word *science* has applied to any system of thought. In ancient times, astrology, necromancy, and medicine all were sciences, though they lacked a common methodology or conceptual basis. It is confusing, however, to retain such a broad meaning of the word. Modern organized science is a system of thought so different from its predecessors that it is necessary to restrict the word *science* to this new system.

Such a restriction affects how we interpret ancient and traditional ideas and practices. Egyptian medicine, for example, is known to us from seven papyruses. The Smith papyrus, from 1700 B.C., is, like the others, a copy of a much older work. Indeed, we know that the text was already ancient in 2500 B.C. because a scribe from that period added the definitions of words that were by then obsolete. One side of the papyrus contains spells and recipes, and the other describes forty-eight surgical cases. Scholars have been much impressed by the matter-of-factness and soberness of these ancient texts and, in particular, by the clinical tone of the Smith papyrus. A renowned historian of science, George Sarton, concluded that "the Smith papyrus, and to a lesser extent the [others], give us a very favorable idea of the medicine, anatomy, and physiology of the Egyptians, and of the scientific outlook that they had obtained at least 2,000 years before Hippocrates" (1960, pp. 47–48).

Sarton has no trouble speaking of Egyptian science because of his broad definition of science:

> What is science? May we not say that whenever the attempt to solve a problem is made methodically, according to a predetermined order or plan, we are witnessing a scientific procedure, we are witnessing the very growth of science? To be sure, early

methods seem childish and weak as compared with ours, but
will the scientists of the year 5000 think as favorably of our
methods as we do ourselves? . . .

Some readers having at the back of their minds the preju-
dice that science is a Greek invention (have not scholars repeated that for centuries?) will insist and say, "That may be science, but not pure science." Why not? . . .

I am sure that not only the Egyptians who had reached the
stage of composing mathematical and medical treatises, but
simpler men, living perhaps thousands of years earlier, were already pure scientists, that is, men moved by such an intense
curiosity that the practical results and immediate fruits of their
research became of secondary importance to them. (1960, pp.
48–49)

But science isn't just the methodical solution of a problem, "according to a predetermined order or plan." It isn't true, as one ninth-
grade science textbook states, that "whenever you observe the world
around you, you are acting like a scientist" (Hurd et al., 1991, p.
6). Every astrologer, homeopathist, Tarot-card reader, witch doc-
tor, or psychoanalyst makes observations and works according to a
predetermined order or plan. To call such work scientific, or even
prescientific, is to miss the essential meaning of science. These nonscientific systems are tradition bound and self-referential, working
within their own closed system of ideas, much as Kuhn claims that
science does. Such nonscientific systems have been around since Paleolithic times. They are as popular today as ever, with major publishers eager to publish accounts of UFO abductions or ancient civilizations on Mars.

Harvard psychiatrist John Mack, for example, will soon be publishing his much publicized study of UFO abductees. His belief that
accounts of UFO abductions present an "authentic and disturbing
mystery" is based on his personal assessment "as a psychiatrist" that
the abductees he interviewed are reliable informants and on his judgment that the details of their narratives are consistent (Mack, 1992).
Yet he admits, when questioned, that the abductees give very different descriptions of their abductions and abductors. A mystic in
search of a mystery, Mack prefers to believe that "we can be 'invaded' or taken over . . . by some other form of being or consciousness" (p. 11), rather than that the abduction stories are based
on vivid violation dreams in which science-fiction imagery has replaced traditional demons.

All nonscientific systems of thought accept intuition, or personal insight, as a valid source of ultimate knowledge. Indeed, as I will argue in the next chapter, the egocentric belief that we can have direct, intuitive knowledge of the external world is inherent in the human condition. Science, on the other hand, is the rejection of this belief, and its replacement with the idea that knowledge of the external world can come only from objective investigation—that is, by methods accessible to all. In this view, science is indeed a very new and significant force in human life and is neither the inevitable outcome of human development nor destined for periodic revolutions. Jacques Monod once called objectivity "the most powerful idea ever to have emerged in the noösphere." The power and recentness of this idea is demonstrated by the fact that so much complete and unified knowledge of the natural world has occurred within the last 1 percent of human existence.

From our present perspective, we can see that most of the currently established scientific knowledge—the stuff that's in the textbooks—is a permanent part of human knowledge. Scientists in the year 5000 will "think as favorably of our methods as we do ourselves" because their work will be firmly based on it. If anything, they may be awed that the twentieth century learned so much using such primitive instruments.

In the interpretation of science presented in this book, objectivity itself is viewed as a peculiar human invention, with its own unique history. Furthermore, objectivity is complete and self-correcting in the sense that it is capable of learning whatever human beings are capable of learning about the external world—it's as far as we can go. This interpretation raises many interesting questions and has some startling implications.

If there is completeness of scientific knowledge, then there are limits to the technologies based on this knowledge. The same knowledge that enabled us to send men to the moon tells us we'll probably never send humans to Mars, and certainly never to a nearby star. In spite of its many prominent boosters, space travel is near—if not at—its practical limit.

Our teaching of scientific thinking, on the other hand, is still at a primitive stage. There isn't any agreement among educators on what to teach or how to teach it, or even if science is really objective. This confusion stems from a failure to appreciate the nature of objectivity. At its root, objectivity is simply respect for the critical opinions of others.

Why is this so difficult? Why has it taken humankind so long to discover it? And why, having discovered it, do we so readily abandon it for cults and mysteries? These are some of the questions we'll try to answer in Chapter 2, where we consider the potential of the human mind for both rational and magical thinking.

2
Mind and Magic

Opinions about the human mind are as numerous as the multitude of anthropologists, biologists, psychologists, and philosophers who have written about it. It's rare, however, for a physicist to venture into such a discussion. We don't dig for ancient bones or study the ways of bees and geese; our research deals with the inanimate world of electrons and protons, far removed from questions of human behavior.

Yet as teachers we confront, on a day-to-day basis, some deep issues concerning the nature of the human mind. With each new freshman class, I again must face the fact that the human mind wasn't designed to study physics.

A Universal Error

I have a toy gun that shoots Ping-Pong balls. If I fire one ball horizontally while at the same time—and from the same height—I drop a second ball, which one will hit the ground first? The universal reply is that the first ball will hit the ground after the second because the first travels a greater distance (Figure 2.1). This was the opinion of Aristotle, and it was universally held until 1632 when Galileo wrote that

> if a perfectly level canon on a tower were fired parallel to the
> horizon, it would not matter whether a small charge or a great

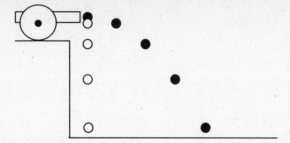

Figure 2.1 A ball fired horizontally from a cannon strikes the ground at the same time as does a second ball dropped from the same height at the instant the cannon is fired.

> one was put in, so that the ball would fall a thousand yards away, or four thousand, or six thousand, or ten thousand, or more; all these shots would require equal times [to hit the ground], and each time would be equal to that which the ball would have taken in going from the mouth of the canon to the ground if it were allowed to fall straight down without any other impulse. (Galilei, 1632/1967, p. 155)

The notion is astonishing. Three hundred and fifty years after Galileo, my students still react with surprise and delight when the ball fired from the toy gun hits the ground simultaneously with the dropped ball. Eighth graders, less impressed by authority, aren't convinced even when they see the demonstration.

This is quite surprising, since men have been throwing rocks, spears, and arrows at one another for thousands of years. You would think it would be common knowledge by now that all missiles thrown horizontally from the same height hit the ground in the same time. Clearly, experience wasn't enough. Galileo used logical analysis, not experience, to determine the true motion of missiles. I am convinced that the two Ping-Pong balls hit the ground at the same time not only because I can make it happen, but also because this behavior is part of a logically consistent theory of motion.

When I first started teaching introductory college physics thirty-some years ago, my unanalyzed assumption was that my students, though ignorant of physics, could think logically. Therefore, my job was to prepare logical lectures, speak slowly, and write big on the blackboard. So confident was I of this approach that I thought I could teach physics "right the first time." This meant including Einstein's relativity in beginning lectures on Newtonian mechanics, which is akin to including calculus in a third-grade class on long division.

The New Math, which was being introduced into schools about that time, was based on the same assumption—that the mind is hard wired for abstract logical thinking. The failure of the New Math in general, and of my own approach in particular, makes it necessary to question that assumption.

Galileo believed that each mind has prior knowledge of the truth and that the teacher's role is to make the student aware of this knowledge: "You yourself also know . . . [it] . . . , no matter how you may pretend not to know it—or give that impression. But I am so handy at picking people's brains that I shall make you confess this in spite of yourself" (Galilei, 1632/1967, p. 145). We no longer believe that our students come to us with the laws of physics buried in their heads. Moreover, experience has shown that they don't have even the prerequisite logical skills to understand the laws when they see them. Both the skills and the knowledge have to be learned together.

Piagetian Psychology

The work of the Swiss psychologist Jean Piaget on the mental development of children provides a model of the human mind that is consistent with the experience of physics teachers (Wadsworth, 1971). Instead of believing that the brain contains preexisting information-processing structures, Piaget sees such structures emerging in stages of advancing complexity. The four Piagetian stages are termed sensorimotor, preoperational, concrete operational, and formal operational. Only those students who have reached the formal operational stage of development are ready for abstract subjects like physics and geometry. And according to Piaget, development from one stage to the next requires the accumulation of relevant experiences.

It is found, for example, that children aren't able to apply logic to verbal problems before they are able to classify and order things. And they can't classify and order before they realize that the number of items in a group doesn't change when the items are rearranged. These abilities don't develop naturally as a child matures, but are learned in a variety of formal and informal ways from older individuals who already possess them. The ability to classify, and all the higher skills that build on it, won't develop unless there are things in the culture, such as laundry, stamps, or baseball cards, for a child to classify.

Piaget's work was mostly done with working-class children in a few public schools in Geneva. Coming from a common background, these children exhibited similar patterns of mental development. Children with a different background would be expected to follow their own unique pattern of development. The earliest cognitive structures are formed from the newborn's first unprejudiced experiences with the world. The next structures, however, are formed from experiences as interpreted by the first structures. The third grow out of the second, and so on in an unbroken progression.

Since many intelligent students are unable to grasp mathematical logic, one must suppose that not all normal developmental sequences lead to this ability. This means that mental growth, unlike physical growth, isn't innate. Given adequate nutrition, an individual grows from infancy to adulthood through physical stages that are precisely programmed by the individual's inherited genes. In contrast, the stages of mental growth aren't immutable or inevitable, but depend strongly on an individual's cultural and social environment. By age three, the disadvantaged child who has been sitting passively in front of a television set all day is a profoundly different human being from the preschooler who has been interacting with peers and teachers in a variety of games and projects of increasing complexity. The passive child may already be missing cognitive structures essential to progression to the stages of concrete and formal operations.

The stage of formal operations, nominally reached by adolescence, is characterized by the ability, absent earlier, to solve word problems, to deal with hypothetical questions, to use several concrete operations simultaneously, to theorize, and so on. Yet in a recent study of the mathematical ability of seventeen-year-olds in the United States, less than 6 percent could solve simple algebra problems (Saltus, 1989). It has been known for some time that most American college freshmen haven't reached the stage of formal operations (Lawson and Renner, 1974). This explains why my early attempts to teach physics as a system of logically interconnected ideas was doomed to failure: My students didn't have the mental structures required to handle such an approach.

Here is a simple question at the boundary between concrete and formal thinking: A wooden block is one inch long on each edge. How many such blocks does it take to make a cube two inches long on each edge?

Many of my college students don't get it, even when the blocks are on a table in front of them. On the other hand, one ten-year-old, told to work it out with pencil and paper, got it quickly. From the boy's joy with his success, I concluded that the problem exercised his current cognitive structures and stimulated the development of higher ones. College students who never developed the structures for dealing with such a problem are unable to see the significance of the problem even when the answer is explained to them. They certainly don't indicate by their enthusiasm that they have just perceived a profound truth about the structure of space. (For the answer to this question, see Figure 2.2.)

Historical Development of Logical Thinking

Everyone accepts that physics must be formally learned. But now, from studies of how students learn (or don't learn) physics, it is clear that the thinking processes themselves must be formally learned. Logical thinking doesn't develop spontaneously from everyday activities. Practically, this means that the development of abstract logical thinking should be the principal goal of science courses at all levels, with the science itself being the vehicle for reaching this goal. Conceptually, this means that logical thinking must have arisen his-

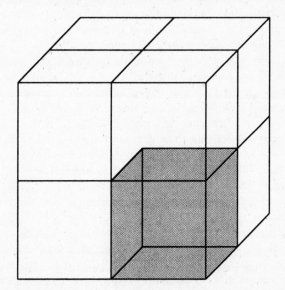

Figure 2.2 It takes eight cubes one inch on a side to make a larger cube two inches on a side.

torically from very special cultural factors, unconnected with science.

It is reasonable to assume that historically, the mentality of humankind has passed through stages of development similar to those of an individual child today. Since every culture has a different history, each has its own hierarchy of mental structures. Logical thinking, as evidenced by the development of deductive mathematics, seems to have been a unique characteristic of Greek civilization. As we'll argue in detail later, this attributes the origin of science to very special cultural factors peculiar to Greece, rather than to some generalized human tendency toward rational thought.

Egocentrism

If there is one universal human characteristic, however, it is a pervasive irrationality based on the egocentric confusion of self and other. At birth, an infant makes no distinction between self and other, between wanting and getting. For the well-cared-for infant, to want is to have, so that an infant's earliest experience blurs the distinction between the internal and the external world. As Piaget put it: "Going back to the starting point in the life of thought we find a protoplasmic consciousness unable to make any distinction between the self and things. . . . There is thus in the beginning neither self nor external world but a *continuum*"(1929/1960, pp. 235–36).

This is the basis of egocentrism, which Piaget has identified as the major obstacle to mental growth. As a child grows and progresses through the Piagetian stages, egocentrism reappears in different forms. In the sensorimotor stage, the infant doesn't distinguish self from others (people or objects). Then, as the child first starts to relate objectively to others, it naturally assumes that these others—whether mother, blanket, or teddy bear—are beings like himself. A child's emotional attachment to physical objects, which is a necessary stage in his ability to eventually form emotional attachments to people, grows out of his egocentric identification with the object.

In the preoperational stage, a child doesn't distinguish her thoughts from the thoughts of others. Since the caring parent often knows before the child does when it is time to go to the toilet, the child naturally believes that her thoughts have affected her parent's behavior. A four-year-old may believe that her thoughts also have the power to affect what happens to others, and so she will feel guilty should misfortune befall someone about whom she has had an angry thought.

Piaget identified this primordial egocentrism as the origin of animism:

> The social and biological factors combine at the beginning of the mental life to ensure an absence of differentiation between the world and the self, whence arise the feelings of participation and the magical mentality which results.
>
> If such is the starting point for the child's consciousness it is easier to realise the origins of animism. (1929/1960, p. 236)

Animism is the attribution of aspects of the self to objects and events. As Piaget emphasized, it isn't a projection of consciousness into animate and inanimate objects but a primitive mentality that has yet to differentiate between the internal and external, between thought and object, between a name and the thing named. All children are metaphysical realists in that they believe words are the thing they name. Even as late as nine years of age, Piaget's Geneva children thought the sun was called "sun" because it is hot or because it is red. The concept of words as nominal representations of objects and the concept of thoughts as subjective mental images develop much later than we might think and then only after years of formal education.

The child's rhyme "Sticks and stones may break my bones / But names can never hurt me" is both a charm against the power of words and an emerging recognition of their nominal nature. Freedom of speech, a rather new idea still not universally practiced, requires a general understanding that words, although they can influence the actions of others, don't in themselves have power. The murderous outrage against Salman Rushdie for his book *The Satanic Verses* can perhaps best be understood as arising from an undifferentiated confusion between the word and the thing it represents. Certainly, there was no danger that this fantastic novel would undermine the faith of a devout Muslim or lead to revolution in Iran. It was the words themselves—and the person who wrote them—that were condemned, not their possible influence.

Although it sounds contradictory, what we call objective thinking is possible only after we come to understand the subjective nature of thought. This is because objectivity requires that we differentiate between the internal world of private thoughts and dreams and the external world that exists apart from us. Indeed, without

such a differentiation there is no external world—or internal consciousness, for that matter—but only an undifferentiated stew of mind
and matter. Thus science—the grand exemplar of the power of human thought—is possible only after it is recognized that thought has
no "real" power. For many, this final break with animistic and
magical thinking has been too high a price to pay for science.

Animism

All children are animistic, and animism continues throughout life
unless strongly controlled by contrary cultural guidance. It is universal among preliterate peoples, who believe that spirits inhabit all
things—animate, inanimate, and supernatural. The prehistoric An-
asazi of the Southwest would chip the rim of a clay pot before placing it in a grave so that the spirit of the pot could escape. But animism is a strong component of modern life as well, from belief in
the power of an amulet to ward off the evil eye to my belief that it
rains only when I forget my umbrella.

An extreme example of such egocentric thinking was the beliefs
of the people of the Dobu Islands, as described by Ruth Benedict
(1934/1989). The Dobu Islands are off the southeastern coast of New
Guinea and are now part of Papua New Guinea. At the periphery
of the Melenesian world, the Dobuans of sixty years ago survived
by growing yams in individual family plots. They believed that
everything was controlled by magic and so that to grow their yams
they must recite the correct yam-growing incantations. These, however, weren't general prayers valid for the entire group, but secret
incantations specific to a family's own yam seeds. These incantations
were jealously guarded and handed down from a man to his wife's
oldest sororal nephew.

The Dobuans further believed that the yam tubers left the ground
each night to wander about and so had to be cajoled back to their
garden each morning. There were specific incantations for this,
however. In addition, there were incantations by which each Dobuan gardener tried to lure his neighbors' wandering tubers to his
garden. An abundance of yams at the end of the growing season
didn't attest to the gardener's gardening skill, but to the power of
his yam-stealing incantations. Consequently, a successful harvest had
to be kept secret in order to avoid retaliation.

The countless incantations and magical charms that a Dobuan
needed weren't shared with the community, but had to be acquired

by purchase, inheritance, or theft. The Dobuan were so preoccupied with acquiring magic charms and spells that would injure a neighbor, cure a disease, seduce a lover, win a wager, and the like that there was little time or inclination to learn objective skills. They were, in fact, a wretched people living on the least desirable islands of Melanesia. Since their survival skills had to have come from somewhere, we must assume that they were the impoverished remnants of a more successful people. They are a stark example of the hold that magical thinking can have on human behavior.

Religion

In the formal operational stage, an individual doesn't distinguish ideas from reality. The adolescent is so overwhelmed by the apparent power of his own thoughts that these thoughts are held to be true because he thinks them. When humankind first reached this same stage, new ideas about religion had the impact of divine revelation, for it isn't self-evident where new ideas come from. Divine inspiration, subconscious, and the right side of the brain are different explanations of an as-yet unsolved mystery.

Magical spells and religious prayers are expressions of the egocentric belief in the omnipotence of one's words and thoughts. As such, the number of different superstitions and religions is as boundless as thought itself. Established religions depend on the egocentrism of their believers while trying to channel it into their particular orthodoxy. But even the most devout will try other supernatural vendors from time to time. It is a wise religion that doesn't look too closely for heretics in its midst, for it will certainly find them.

A spell or magical ritual differs from prayer by being independent of the motivation or legitimacy of the supplicant and by its need to be recited or executed perfectly. The power is in the words or actions themselves. The agent to which the spell is directed acts automatically if the words are right, regardless of the supplicant's intentions or state of grace. Magic is analogous to science in that both believe that the external world is controlled by impersonal forces. It differs from science by the belief that the forces of nature can be cajoled by words or thoughts.

A prayer or offering, on the other hand, is made to a supernatural being who has the power to grant or reject the supplicant's wish. Although the correct words of a prayer also are important, this correctness is more for sentimental than magical reasons; slight errors in form don't invalidate a prayer, though they may weaken it. In religion, the supplicant must develop a desirable personal relationship with the supernatural in order to receive its benefits.

Religion is a development of the hierarchical state, which first emerged in the Middle East some 10,000 years ago. There, animistic relations with the spirits of plants, animals, and places gradually gave way to the religious worship of gods with human attributes that closely resembled those of the ruler and his nobles. This egocentric confusion of the structure of the supernatural world with that of the state probably occurred when the intellectual elite of the time—the priests, astrologers, and physicians—were first reaching the stage of formal operational thinking. The egocentric belief in the blinding truth of one's thought, which is characteristic of this stage, became the source of religious zeal and conviction.

Individuals with a gift for oratory became prophets and preachers, exhorting whomever would listen with their inspired interpretations of the divine. Some were sincere religious reformers; some were charlatans; and some were tormented schizophrenics. Echoes of madness resound in the lament of the prophet Jeremiah:

> I am a laughing-stock all day long, they all make fun of me.
> For whenever I speak, I have to howl and proclaim, "Violence
> and ruin!" For me, Yahweh's word has been the cause of insult
> and derision all day long. I would say to myself, "I will not
> think about him, I will not speak in his name any more," but
> then there seemed to be a fire burning in my heart, imprisoned
> in my bones. (20:7–9)

All the spiritual leaders of the literate world—Moses, Buddha, Jesus, Muhammad, Smith—have based their teachings on personal visions they believed to be of supernatural origin.

Peoples sometimes use extraordinary means to obtain visions. It was a common practice among Native North Americans (with the exception of the Pueblo Indians) to seek visions through solitary vigils, fasting, and even self-mutilation. The insights that came to a man at this time were considered to be of supernatural origin, thus giving his ideas special authority. The content of these insights necessarily dealt with the particular concerns of the visionary's tribal society and usually consisted of some minute change in the customary rituals. Today visionaries are as likely to see the face of an extraterrestrial as the face of God.

Egocentrism is clearly a dominant component of the human personality. It is the basis of superstition, magic, witchcraft, and religion, as well as leadership, artistry, innovation, and, indeed, any thought or activity that an individual believes has meaning. Egocentrism, as used here, doesn't imply egoism or selfishness. In fact, empathy and altruism are expressions of the egocentric idea that the self exists in others. Egocentrism is subjective thinking in all its aspects, and it is a basic aspect of the human condition.

Science and Magic

Every culture deals with its environment with a combination of naturalistic and magical–religious practices. It has been customary for anthropologists to use the word *science* to refer to the technological and naturalistic knowledge of any culture. As William Howells wrote:

> Religion therefore may be said to complement science. For anthropological purposes . . . . I should define science simply as the understanding of the correct explanation or principle for anything at all. If this is so, then the simplest technical exercise, like baiting a hook so that a fish will want to bite it, is scientific, while putting a spell on the same hook and bait to make the fish bite, regardless of his appetite, is religious. Thus people solve their problems either by science or by religion, since, if science hasn't an answer, religion has it. (Quoted in Hsu, 1983, p. 6)

For our purposes, however, the critical question is whether the distinction between religion and magic, on the one hand, and science, on the other, is understood by the culture using them. Malinowski firmly believed that they are: "As a matter of empirical fact, the body of rational knowledge and the body of magical lore are incorporated each in a different tradition, in a different social setting, and in a different type of activity, and all these differences are clearly recognized by savages" (quoted in Hsu, 1983, p. 5). The anthropologist Francis Hsu disagrees. From his observations of the actions taken by the inhabitants of a Chinese hamlet during a cholera epidemic in 1942, he concluded that most people don't distinguish between their magical–religious manipulations and their reality-based practices:

> In search of comparative material, I looked to the world beyond China. What I saw, from the South Seas and Africa to Indian and white America, was that, despite industrial and

technological advances and Western influences, the line between magico-religion and the empirical–scientific remains unclear in the minds of the majority. (Hsu, 1983, pp. xv-xvi)

I have reached a similar conclusion from years of teaching normal American college students. Most don't have the critical thinking skills needed to distinguish the fanciful claims of astrology from the extraordinary claims of astronomy. Indeed, few people have, which explains the wide acceptance in our technologically advanced culture of every manner of superstition. A study at a Canadian university found that a majority of the students believed in astrology, extrasensory perception, and reincarnation (Gray, 1987), and a Roper poll found that 11 percent of Americans have seen a ghost (Chandler, 1992). Spokespersons for astrology, spiritualism, demonology, UFOlogy, and other occult phenomena have far more access to and acceptability in the popular media than do the scientists who try to refute them. This is partly because occultism is a form of entertainment and rational skeptics aren't very amusing. But basically, the popularity of the occult arises from the deep-seated egocentric need to believe that one is somehow linked in a personal way with the external world.

For the most part, occultists wrap their work in the mantle of science. Thus Joan Quigley, Nancy Reagan's clairvoyant adviser, calls herself a "serious scientific astrologer" (Plummer, 1988, p. 9) and V. James DeMeo, a modern proponent of Wilhelm Reich's idea that the orgasm is a cosmic force, claims that "every phase of Reich's orgone theory was derived experimentally" (Gardner, 1988, p. 29).

To this day, science is alien to most people, and not just because of its technical difficulty. Many people who have mastered the technicalities still refuse to accept science's banishment of the subjective. Science has a string of intellectual critics who would make it into something else: something more intuitive, subjective, direct; something that would admit knowledge based on an individual's sensitivity to nature; something in which egocentrism would still play a role. This hostility to science, in the face of its obvious triumphs and benefits, is additional evidence that it is something outside the mainstream of human development, perhaps a fluke.

Unfortunately, the notion of science as a fluke runs counter to the thinking of most respected scientists, who are inclined to explain any phenomenon, science included, as the predictable consequence of causal forces. This approach has been very successful for geology and evolution, in the face of strong opposition from the creationists, and so scientists may be forgiven for thinking that my fluke is a flake.

Chaos and History

But recently, from studies as diverse as Newtonian mechanics and early Cambrian arthropods, there has been a major reorientation in thinking about causality and predictability. Specifically, it is now understood that there can be causality without predictability, even in simple mechanical systems. Happily this is one of those ideas that, though obscure to scientists, is clear to lay people.

Your life is a causal system in the sense that what happens to you at one instant is caused by what happened to you (and everybody and everything else) just an instant earlier. Nevertheless, your life is unpredictable because a very small change in your circumstances at any instant can have very large consequences later in life. You do or don't get hit by the drunk driver; your eye does or doesn't catch the eye of the attractive stranger; you do or you don't run into an old friend with an important proposition. Minute changes in your life at any instant can profoundly change your life's trajectory, which is what makes your life interesting and unpredictable. As obvious as it is, it wasn't taken seriously until blessed by the insights of physics.

In Newtonian mechanics, the motion of a mechanical system, such as that of the earth about the sun, is determined by mathematical equations. The solution of the equations is a relationship that describes where, given their current positions, the objects of the system will be in the future. For this reason, the notions of causality and predictability have been thought of as synonymous. With the advent of the computer, however, it has been found that some systems, despite evolving in time in a totally causal manner, are unpredictable because minute changes at one time produce great changes later on. Such systems are said to be chaotic.

The double pendulum is a simple and dramatic example of a system that exhibits chaos (Shinbrot et al., 1992). Although I am a theoretical physicist, I have learned from my experimental colleagues how easy it is to build a double pendulum: Ask our machinist Dick Ahlquist to make one. Since 1989, Ahlquist has built exhibition-quality double pendulums for our department, for the Boston Museum of Science, and for loan to middle schools in the area (Cromer, Zahopoulos, and Silevitch, 1992). They are fascinating to watch, for even though they are a simple mechanical device, they display some of the spontaneity of life itself.

A double pendulum has two arms that rotate around fine bearings (Figure 2.3). The upper arm is made of two nine-inch-long aluminum bars mounted on bearings fixed to a rigid stand. The lower arm is mounted on a bearing fixed to the free ends of the upper arm. The bars of the upper arm are spaced just far enough apart to allow the lower arm, made of one eight-inch-long aluminum bar, to rotate between them. When the upper arm is lifted a small amount and released, the two arms swing back and forth in a regular manner. Given any small initial positions of the two arms, their positions at any later time are given by the solutions of Newton's equations. The motion is causal and predictable.

But if the upper arm is lifted a large amount and released, the situation is very different. The small arm may swing back and forth as before and then suddenly make a few complete clockwise rotations, followed by one counterclockwise rotation, followed by several back-and-forth swings, and so on in an apparently spontaneous and unpredictable fashion. Newton's equations still apply, and their solutions show similar patterns of swings and rotations, but given any large initial positions of the two arms, their positions at any

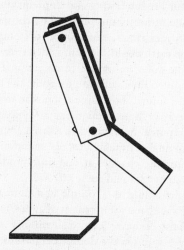

Figure 2.3 A double pendulum displays chaotic motion when it is released from an angle large enough to allow the lower arm to swing around in a complete circle.

later time can't be determined by solving Newton's equations. The motion is causal and unpredictable.

Chaos occurs once the lower arm is able to make complete revolutions, because each time it approaches the top of its arc, it will either continue around or swing back in the opposite direction. Which it does depends on its speed and on the speed and position of the upper arm at that instant. Slight changes in the motion of the upper arm can cause the lower arm to go in opposite directions. Either way it goes, the lower arm will soon reach the top of its arc again, and again may go in two opposite directions. As this continues, the motion of the double pendulum becomes ever more convoluted and unpredictable. Two Ahlquist pendulums, built of identical stock and released as carefully as possible from the same positions, will soon be swinging very differently because the slightest difference in the pendulums or their starting position will eventually cause the small arm of one to complete a rotation, while the small arm of the other swings in the opposite direction. Thereafter, the two motions go their separate ways. No two motions are ever the same. Each motion of the double pendulum, just like each life, is episodic, with a sequence of "defining moments" whose outcomes are the unique history of that motion.

Biological evolution can be looked at in much the same way. The evolutionary history of any species, including our own, is a sequence of critical junctions at which speciation could go along alternative routes or extinction could occur. Looking back, we may be able to identify some of these junctions, but we can't look forward and predict them. Each species has its own unique history.

This brings us to the arthropods of the early Cambrian period. These small creatures, mostly less than an inch (25 mm) long, swam in shallow seas 530 million years ago. Although all traces of most of them were obliterated ages ago, there is a remarkable rock formation in British Columbia, called the Burgess Shale, that contains the fossilized remains of many of these delicate creatures. Meticulously, with needle and microscope, paleontologists have been reconstructing their architecture from the filmlike traces they left in ancient mud. An artist's renderings of how they may have looked in life fill Stephen Gould's (1989) wonderful book about life in the Burgess Shale. What is so fascinating about this life is its great diversity. Whereas today all 750,000 known species of arthropods belong to one of three major groups (insects and millipedes, spiders and mites, and crustaceans), the 50 arthropod species found in the Burgess Shale represent more than 20 radically different groups, all equally cute. Through all the twists and turns of fate, the descendents of only three of these groups survived to this day. If, as Gould puts it, the tape of history were to be rerun, it is unlikely the same groups would survive, since history, like the double pendulum, never repeats itself.

Looking back from the present high point of scientific and technological development, it is tempting to suppose that evolution, however convoluted its path, must sooner or later arrive here. Since technology is such a great thing to have, must it not develop eventually on any planet with evolving life forms? Not necessarily.

Science is the end product of two unique histories: the biological history in which human beings evolved from anthropoid ancestors, and the human history of the discovery of objective thinking and its application to the study of nature. The second history wouldn't be possible if the first didn't endow us with the capacity of abstract thought. But why do we have this capacity? What were the specific environmental factors that selected for high intelligence? Were they the same ones that first caused the hominid line to diverge from the ape line? If so, then in a sense, science was the destiny of our line. If not—if the hominid line from which we are descended arose from extraneous factors having nothing to do with heavy thinking—then human intelligence itself is a fluke, an unpredictable event along the chaotic trajectory of life. Maybe we got to where we are by such a tortuous path that its likelihood of being repeated elsewhere is very small.

To get some insight into these questions, we'll review in Chapter 3 the current knowledge of human evolution and what it tells us about who we are and why we are the way we are. In particular, we are looking for the environmental factors that favored an increased brain size.

3

From Apes to Agriculture

The Origin of Species (Darwin, 1859/1979) is the most in-fluential book since Newton's *Principia*. Based in part on observa-tions made by Charles Darwin (1809–1882) during his five-year round-the-world voyage aboard HMS *Beagle* and in part on his in-vestigations of the methods of animal breeders near his country home in Kent, England, it is an epic synthesis of an immense array of facts into a thesis of elegant simplicity and penetrating depth. The great controversy that this thesis engendered, because of its incompatibil-ity with the biblical creation myths, continues to our day (Irvine, 1959; Kehoe, 1987; Stone, 1980).

Evolution and Natural Selection

Origin actually asserts two theses. The first is the principle of evo-lution, which states that each existing species isn't a separate crea-tion, but is the descendant of some previously existing species. The principle of evolution explains, and is supported by, the fossil re-cord, that vast assembly of fossilized bones and shells of millions of extinct species that lie, layer on layer, in the sedimentary rocks de-posited over hundreds of millions of years. These fossils can be ar-ranged into ancestral trees that connect earlier species to later ones in graduated sequences.

The second thesis is the principle of natural selection, which is a statement about the cause of evolution. Darwin knew that in every species many more eggs and young are produced than ever live to reproduce themselves. Influenced by the ideas of Thomas Malthus (1766–1834) concerning population growth and food supply, Darwin recognized that the excessive reproductive capacity of plants and animals creates fierce competition for survival. He proposed that this competition would evolve new species by naturally selecting the better-adapted individuals over the less-adapted ones, just as breeders develop new varieties of domestic plants and animals by artificially selecting those individuals with the more desirable characteristics.

Both artificial and natural selection are based on two facts: First, there is a great variation of characteristics among the individuals of any species, and second, these variations are usually inherited by the progeny of an individual. Consequently, if in a population of cats, only the longer-haired cats and their longer-haired descendants are allowed to breed, after a number of generations a variety of long-haired cats will be produced. Extrapolating from this, Darwin suggested that natural selection, working on small variations over millions of generations, had evolved not just new varieties, but also new species, families, classes, and phyla.

Darwin had abundant, but not overwhelming, evidence to support his radical new concept of the natural order. In its original form, Darwinism had substantial defects. For example, it lacked any understanding of how variations in organisms arose and how they were inherited. The genetic theory of Gregor Mendel (1822–1884) didn't become widely known until forty years after the publication of *Origin,* and not until the work of Crick and Watson in the 1950s was the molecular basis of Mendel's genes finally deciphered.

Yet even today there is no theory of evolution, no set of rules that says under what conditions this or that will happen. Natural selection is a very general explanation of how evolution occurs, but not when or why. As with any historical process, so many minuscule factors can affect the outcome that prediction is impossible. Two worlds, with the same plants, animals, and environment, will over time evolve radically different species (Gould, 1989). The trajectory of evolution is chaotic. Looking back in time, evolutionists can try to infer what environmental factors influenced the development of certain species, but this is little more than speculation. Too many factors are involved ever to be sure which the instigating one was.

In the case of human evolution, the situation is even more speculative because the fossil evidence is so spotty. Every new discovery has the potential for radically changing our views on the subject. This is a very incomplete science in which the most basic question—What were the selective pressures that favored human intelligence?—remains unanswered. That is, we don't know what natural selection was doing when it invented thinking.

Nevertheless, a review of what is generally known about human evolution is useful, if only to dispel the notion that it followed some smooth inevitable path. Human evolution is often pictured as a steady march toward higher intelligence, but this is misleading. Each step took millions of years, and the environmental factors at each junction were very different. The outcome of evolution, being the result of many independent fortuitous factors, is essentially random.

Extinction of the Dinosaurs

The most fortuitous event of all may have been the asteroid that struck the earth 65 million years ago. Evidence is mounting that this asteroid struck the north coast of the Yucatán peninsula, leaving a subsurface circular crater 110 miles (180 kilometers) in diameter (Kerr, 1992). Using sophisticated radiometric-dating techniques, the date of impact has been found to be exactly 65 million years ago (give or take 100,000 years), coinciding as exactly as these things go with the end of the Cretaceous period and the extinction of the dinosaurs (Swisher et al., 1992).

Before this event, the dinosaurs were an extraordinarily successful group of animals, dominating the earth and sky and venturing into the oceans as well. The current view is that they were quick, warm-blooded animals that—in some cases, anyway—cared for their young and evolved social organizations (Morell, 1987). Although they didn't develop the higher intelligence required to make tools, in their almost 200 million years on earth, they did evolve most of the other animal abilities: running, flying, swimming, grazing, preying, and so forth.

The first mammals were inconspicuous creatures that lived in the interstices of the dinosaurs' world. The first primates evolved from small tree-dwelling insectivores sometime in the late Cretaceous period (perhaps 80 million years ago). The extinction of the dinosaurs left many vacant ecological niches, which the primates, and other surviving mammals, evolved to fill.

Monkeys evolved some 15 million years after the extinction, and the apes, 30 million years after that (about 20 million years ago) (Macdonald, 1984). For a time, the apes were an abundant and diverse group, with large populations and many species, adapted to the tropical forests that then covered Africa. But adverse climatic changes over the last 20 million years resulted in the extinction of all but a few species. And even though human expansion is accelerating the extinction of the remaining apes in Africa and Borneo, extinction has always been a part of the natural order. Old species pass away, making way for new ones, just as people do. On the average, a vertebrate species exists for 2 million to 5 million years (Gould, 1980). Extinction is as necessary for evolution as death is for life.

When a tree dies in a tropical forest, the canopy opens a bit, allowing a beam of sunlight to illuminate the dark forest floor. Here the plants of the forest engage in fierce competition for the moisture and sunlight in this tiny space. Trees—specialists in capturing sunlight from their rivals by growing to great heights—can win the competition as long as there is sufficient moisture to irrigate their extensive vascular systems. But as the climate of Africa became drier, those plants able to survive drought began to win this space more and more often.

A relatively new order of plant, the grasses (poales), emerged as the key competitor of trees. Grasses are plants with fibrous roots and a jointed stem. The leaves grows alternately on opposite sides of the stem, each joint producing one leaf whose base sheaths the stem. Grasses also have underground stems, called rhizomes, that store food and produce new plants asexually. Descended from flowering plants pollinated by insects, grasses reverted to a reduced open flower that is pollinated by the wind. Evolution has no directionality. Natural selection is about survival, not about the development of complex forms.

Modern grasses evolved perhaps 50 million years ago, becoming abundant 30 million years ago. Because of their rhizomes, they can survive drought, fire, and heavy grazing. Once established in a sunlit space on the forest floor, grass can crowd out the seedling of the mightiest tree. Perhaps as the African climate became drier, grass gradually replaced trees as they died, until the woodlands became grasslands. More likely, forest fires, which started during unusually long droughts, destroyed large tracts of forest at one time. With the first rainfall, grass would spring up from wind-blown seeds, and in a few weeks a woodland would be converted to a grassland.

The First Hominids

It is in this environment of dwindling forests and expanding savannas that the first hominids are believed to have emerged. Perhaps there was a period when small wooded islands dotted the grasslands, providing habitat for an ape that could quickly run from one island to another. Or perhaps apes, living on the edge of great forests, were forced by population pressure to spend more and more time out on the grassland in search of food. Again, perhaps a single event, such as a great fire, isolated one troop of apes in a small surviving wooded island, far from the continuous forest. There natural selection, acting on this isolated population as it did on the founding finches of the Galápagos Islands, developed a new species adapted to life in a new environment.

The ability of grass to withstand repeated cutting, which makes it ideal for ornamental lawns, also makes it a continually renewable food supply for grazing animals. Modern hoofed mammals evolved during the Miocene period (26 million to 7 million years ago) in response to the abundant food supply and open spaces that the expanding grasslands offered. The great carnivores, in turn, evolved in response to the abundant prey that these enormous herds provided. And into this scene walked bands of displaced hominids.

The earliest hominids probably spent much of their time in trees, briefly venturing out on the savanna in search of food or to reach another wooded area. Only after millions of years, as the forests vanished, did they become fully committed to the ground. Bipedalism is, then, an adaptation of necessity. But any adaptation, once established, changes the environmental position of a species. The new position has unforeseen consequences, to which the species must make further adaptations.

One of the oldest hominid fossils is the 3.5-million-year-old partial skeleton of a female hominid called Lucy, discovered in 1974 by Donald Johanson and Thomas Gray in the Hadar region of Ethiopia, and analyzed in 1978 by Johanson and Timothy White (Johanson and Edey, 1981). With the left pelvis and femur (thigh bone) intact, the femur can be seen to angle toward the midline of the body, as in humans. This is a clear indication of fully developed upright walking, or bipedalism, since it allows the walker to place each foot directly under the body's center of gravity. In apes, the femur is nearly parallel to the body's midline; while walking upright, an ape places each foot under its hip, forcing it to waddle from side to side. You can imitate this by taking a few steps with your knees about one foot apart. This forces you to waddle from side to side to keep your center of gravity over a constantly shifting base of support.

Studies of the molecular differences in the protein albumin found in apes and humans indicate that the human line diverged from the ape line about 5 million years ago. DNA studies indicate an earlier date, from 7 million to 10 million years ago (Lewin, 1984). If it turns out that the later date is correct, then the Lucy discovery is consistent with the belief that bipedalism was an early step, so to speak, in the divergence of hominids from apes. However, if this divergence was at a much earlier period, there would have been time for many other critical adaptations to have preceded bipedalism. In any event, because of the apelike braincases of the Hadar fossils, it is fairly certain that bipedalism preceded increased intelligence in human evolution.

Anthropologists still argue about why bipedalism developed: Was it to free the hands in order to make tools, or was it to allow hominids to travel great distances to scavenge for food on the savanna? Many regard bipedalism as only a minor adaptation to the dramatic changes taking place in the hominid environment. As the biologists Henry McHenry and Peter Rodman stated, bipedalism is "an ape's way of living where an ape could not live" (Lewin, 1987b, p. 969).

Lucy and other bone fragments found at Hadar are considered by Johanson and White to constitute a newly discovered species, which they named *Australopithecus afarensis.* It is earlier than, and so presumably ancestral to, the later australopithecines: *A. africanus* and *A. robustus.* These species originated some 2 million to 3 million years ago and were extinct 1 million years ago. *Australopithecus africanus* and *A. robustus* aren't ancestral to our own *Homo* line, which branched off just before or just after *A. afarensis.* There is considerable controversy about this—that is, about whether or not *A. afarensis* is ancestral to *Homo,* which can't be resolved without further fossil finds. However, even if the particular bones at Hadar aren't ancestral to *Homo,* some slightly earlier bipedal ape certainly was. Thus, whatever the case, the Hadar fossils show that bipedalism evolved at least 1 million years before the use of stone tools.

Homo habilis

The earliest *Homo* fossils date from 2.4 million years ago and are of a species called *Homo habilis,* discovered in East Africa by Louis and Mary Leakey in 1959. *Homo habilis* had a brain capacity of 0.75 liter (compared with 0.5 liter for australopithecines and modern chimpanzees and 1.4 liters for humans), long apelike arms, a higher and rounder cranium, and a flatter face than the australopithecines, but, like the australopithecines, was only about four feet tall. The earliest stone implements date from around 2.4 million years ago and are from East Africa. Similar implements have been found with 1.8-million-year-old *H. habilis* fossils in East Africa, so it has been assumed that toolmaking originated with this species. This assumption has been confirmed by the recent dating of a small *H. habilis* skull fragment at 2.4 million years (Gibbons, 1992a).

It is toolmaking, rather than bipedalism, that distinguishes humans from the animals. The australopithecines looked and behaved like bipedal apes, whereas *H. habilis,* though similar in appearance, would, with half a pint more brain, have behaved in strikingly more human ways. Any activity as complex as toolmaking, which must be taught from generation to generation, requires the prior development of a still more basic human distinction—culture.

Because of the perishability of artifacts made of bone, wood, and hide, our knowledge of ancient toolmaking is limited almost exclusively to stone implements. It is almost certain that these were but a small fraction of the material culture of *H. habilis,* who probably used the stone tools to cut, carve, pound, and scrape the softer materials. The earliest stone tools constituted a single cultural style, called the Oldowan industry, that persisted unchanged for over 1 million years. Oldowan tools were crudely shaped stones, prepared by a few blows with a hammer stone.

The period in which the Oldowan industry began must have been a very innovative one. Soon afterward, however, some degree of cultural rigidity set in, at least as far as stonework is concerned, because the tools remained unchanged for so long. Of course, *H. habilis* was only half a pint ahead of the apes in brain capacity, so perhaps a little slowness is to be expected. And, too, there could have been developments in wood and bone work that are forever unknowable to us. But still, it remains a fact that a creature that was innovative enough to develop stone tools didn't improve on them for 50,000 generations.

The Oldowan stone-tool industry demonstrates that from the beginning of the *Homo* line, over 2 million years ago, our ancestors were culture-bound creatures. This means that their environment was no longer exclusively that of the natural world, but was that of their increasingly complex cultural world as well. Henceforth, evolution would have to respond to this second environment as well as to the first. And since the cultural world is *Homo*-made, *Homo*'s future biological evolution would, in large part, be adapting to itself.

Social and Breeding Patterns

In considering what evolutionary forces were critical to human evolution, we should remember that males and females often have very different goals. In most mammalian species, males serve mainly as impregnators of the females, having little or no role in the rearing or protection of the young. Females try to raise as many infants as possible, whereas males try to impregnate as many females as possible. Because a successful male can leave far more progeny than a successful female can, evolution can be driven by male rivalry. That is, what's good for the best impregnators is passed on to posterity, for better or worse.

In most mammalian species, the males are larger than the females. Natural selection has favored the larger males not because they can help defend helpless mothers and infants, but because they can outfight smaller males for access to females. Mammalian males often spend an inordinate amount of time and energy fighting with rival males, in contrast with male birds that, once mated, devote most of their time and energy to caring and defending their young.

Primates exhibit a variety of breeding patterns (Foley and Lee, 1989; Macdonald, 1984). Gibbons are monogamous, forming lifelong pair-bonds. Gorillas, on the other hand, are polygynous, a single dominant male maintaining a harem of females and their offspring. Chimpanzees, who with the gorillas are our closest living relatives,[1] form groups of unrelated females monopolized by a group of related males. The size difference between males and females is greater in polygynous species than in monogamous ones, because polygynous males must defend their harems from solitary males. Male gorillas weigh twice as much as females; human males are about 50 percent heavier than females; and male and female gibbons are about equal in weight.

The 3.5-million-year-old Hadar fossils of *A. afarensis* contain bones from individuals of very different sizes. The individual called Lucy was only 3.5 feet tall and weighed about 55 pounds, whereas other bones are from individuals 5.5 feet tall and weighing 110 pounds. This seems to indicate that *A. afarensis* had the extreme size difference usually associated with a polygynous social structure. Alternatively, some anthropologists see these size differences as indicating that the Hadar fossils are of two different australopithecine species.

The question of the early hominids' breeding patterns is critical to an understanding of human evolution. Gorillas and chimpanzees are polygynous; *A. afarensis* probably was polygynous; and *H. sapiens* most certainly is. Because these data points surround the early *Homo* species, *H. habilis* and *H. erectus,* it is a reasonable interpolation to assume that they were polygynous also.

What led our first *Homo* ancestors to develop larger brains than their forebears had is not known. Suggestions that it had to do with their changing from fruit eaters to hunters or scavengers are unconvincing, since many other species hunt and scavenge without having undergone such an explosive growth in mental capacity. Furthermore, studies of the markings on fossil teeth indicate that meat didn't become a major part of the *Homo* diet until much later.

What, after all, was so complex about the environment of *H. habilis* that required it to have a brain larger than that of any other creature? Because the brain size would have had to increase before the first tools could be made, it seems reasonable to assume that this increase was in response to an existing need. Since we have no facts to guide us at this point, I will engage in a little speculation.

Imagine that *H. habilis*'s ancestors lived, as gorillas do, in troops containing one male and several females and their offspring. Then, because of increased predation from carnivores or because of competition among the hominids themselves, it became advantageous to reduce male rivalry to the point that several younger males, perhaps sons, could live with their females in the same troop with a dominant male. Such a change, once started, would apply immediate and intense selective pressure on the species, as single-male troops too stupid to cooperate would be raided by multimale troops. The stupid males would be killed, and their females allocated among the victors. In this scenario, there would be continual intergroup fighting to capture external females so that the males within a group wouldn't be tempted to steal females from one another.

The details of what happened 3 million years ago are unknowable, and speculations about early hominid social behavior are untestable. Still, the hypothesis that the *Homo* brain grew to allow the learning of more sophisticated behavioral patterns has merit. In the first place, it allows for the continuous evolution of distinctive human characteristics from well-known animal behavior. Male chimpanzees, for example, are known to congregate in gangs and to attack isolated males from neighboring communities. The deaths from these attacks are not infrequent, amounting to perhaps one death per year per community. These communities also have complex hierarchical social structures with which each individual must learn to interact in subtle ways. Failure to get along can mean failure to breed, especially for a male (de Waal, 1989; Goodall, 1986; Macdonald, 1984; Small, 1990).

In the second place, this hypothesis goes a long way in resolving the nature–nurture dilemma. The enlarged brain of *H. habilis* enabled it to voluntarily adopt behavior patterns that enhanced its survival. Some of these patterns, such as the male dominance of females, may have originated over 3 million years ago and so form a part of our earliest culture. They aren't obligatory, however, and a brain large enough to have adopted them when they were needed is presumably large enough to abandon them when circumstances require it. Indeed, many cultures, such as the Pueblo Indians of the southwestern United States, long ago adopted a gentler matriarchal social order.

Whatever the origin of *H. habilis*'s increased mental abilities, we know that with it he learned how to sharpen one stone by chipping it with another. It isn't clear what these chipped stones were used for. Studies of the microwear patterns on fossil teeth show that the diet of *H. habilis* was similar to that of modern chimpanzees—that is, mostly fruits. *Homo habilis* wasn't a hunter, though he may have obtained some meat by scavenging. For that purpose, a sharp stone would have been useful in cutting through the thick hide of a dead zebra or gazelle.

The Mental Ability of Apes and Humans

The mental ability needed to discover that striking one stone with another produces an implement that can cut hide isn't much greater than that of modern chimpanzees. In the wild, chimpanzees use blades of grass to get ants out of anthills (Goodall, 1986), and under experimental conditions, one chimpanzee discovered that by putting two bamboo sticks together he could retrieve bananas placed beyond the reach of one stick. In the classic experiment described by Köhler (1927/1956), the chimp Sultan was given two bamboo sticks, one wider than the other. Some bananas were placed on the ground outside his cage, more than one stick length away. At first, Sultan tried to reach the bananas with one or the other stick by putting his shoulder through the bars. Then he dragged a box to the bars, a particularly bad move, as the box got in his way.

> Immediately afterwards . . . he pushes one of the sticks out as far as it will go; then takes the second, and with it pokes the first one cautiously toward the objective, pushing it carefully from the nearer end and then slowly urging it toward the fruit. This does not always succeed, but if he has got pretty close in this way, he takes even greater precaution; he pushes very gently, watches the movements of the stick lying on the ground, and actually touches the objective with its tip. (Köhler, 1927/1956, p. 126)

Of course, he couldn't retrieve either the banana or the outside stick in this way, and the observer frequently had to return the stick to him. The experiment was stopped after an hour. Sultan then played carelessly with the two sticks for a few minutes, and finding himself holding one stick in each hand in such a way that they lined up, he pushed the thinner one into the thicker, jumped up, ran to the bars, and began to reach for the bananas with the double stick. The sticks kept coming apart, but Sultan persisted until he had retrieved all the bananas.

What is significant about this episode is that Sultan connected his current activity with his previous one. Similarly, to discover the relation between striking stones and cutting hide (or whatever the sharp stone was first used for) requires the ability to remember one thing while doing another.

Psychologists distinguish between long-term and short-term, or working, memory (Waldrop, 1987). Long-term memory is where all our permanent memories are stored, our knowledge of language, faces, facts, procedures, and the like. It is analogous to the memory of a computer or a bank of file cabinets, except that the information in long-term memory overlaps and interconnects. Memories aren't sharp and distinct, like those in a file cabinet, but blurred and merged. Long-term memories are believed to be stored in the network of connections among the neurons of the brain. This network can hold a virtually unlimited number of these fuzzy memories. As the hominid brain increased in size, so did the size of its long-term memory, but it was so large to begin with that it might not have made the most significant difference in mental ability.

Working memory, on the other hand, is much more limited. This is the memory we use while processing information coming to us from the outside world and from long-term memory. It is absolutely essential to problem solving. Even a bee must be able to compare current visible information with stored information in order to return to a previously visited patch of flowers. Thus at the simplest level, a brain needs to be able to handle two "chunks" of memory at the same time. Of course, the complexity of these chunks increases with the complexity of the brain. The random numbers 3, 2, 7, 1 may occupy four chunks of working memory, whereas 1,000,000 occupies only one, because "1 million" is a meaningful unit.

The two-chunk working memory of *H. habilis* was probably more advanced than that of the chimpanzee, in that it could handle chunks with greater informational content. The working memory of modern humans is even more advanced, with an estimated capacity of about six chunks. Still, this isn't enough to enable most people to multiply 24 by 64 in their heads. Only with the invention of writing, just 5,000 years ago, were human beings able to transcend the limitations of their working memory.

Nothing is known about how working memory is handled by the brain. Does it increase automatically with the size of the brain, or does it require new patterns of neural connections? Can the size of short-term memory be increased by training? Whatever the case, the effective use of short-term memory to solve problems does require training and practice.

And this perhaps is the key to understanding the human mind. The vast potential of the human brain is realized only when, through experience and education, it has been exercised and informed. *Homo habilis,* being, we believe, the first real toolmaker, was self-taught. From this point of view, the crude fireless Oldowan culture was a remarkable achievement, requiring the discovery not just of stone tools, but also of the hidden potential of the *Homo* brain. Even more remarkable, perhaps, is that having made this discovery, the culture stagnated for a million years. Why? It's possible that *H. habilis* had reached its intellectual limit, but I doubt it. I tend to think that the small *H. habilis* population, concentrated in central and eastern Africa, had a single culture that, once established, became ossified and incapable of change. After all, cultural conservatism is only moderately reduced in *H. habilis*'s less limited descendants.

Homo erectus

About 1.6 million years ago, *H. habilis* was replaced by *H. erectus,* a *Homo* species that continued the tendency toward larger brains. The earliest *H. erectus* fossils had a 0.85-liter brain capacity, only slightly larger than that of *H. habilis,* but the latest fossils, from 200,000 years ago, had a 1.1-liter capacity, approaching the 1.4-liter capacity of modern humans. *Homo erectus* had a flatter face than *H. habilis,* but still no chin and, surprisingly, a very prominent browridge.

The first appearance of *H. erectus* coincides with the appearance of a new stone-tool culture, called the Acheulian industry, characterized by the bifacial hand ax. This tool was more fully shaped than any Oldowan tool. Flakes were removed from both sides of a stone to produce two sharp edges coming to a point at one end. The other end was rounded so that it could be gripped firmly in one hand. Otherwise, the Acheulian stone industry wasn't much different from the Oldowan stone industry.

Nothing, of course, can be said about the corresponding wood, hide, and bone industries, except that because no notches were made on the stone tools for latching to wood, we know that there were no stone-tipped spears or arrows. The stone tools may have been used to sharpen the points on wooden spears and arrows, as is still done by some hunter-gatherer peoples today. Studies of the wear on stone tools from 1.5 million years ago indicate that they had been used to cut wood.

Gradually the primate foraging economy changed to a hunting-gathering economy. In a foraging economy, a group travels together, each individual feeding itself on the vegetation it encounters. In a hunting-gathering economy, males leave the group to hunt or scavenge meat, while females stay in camp to care for their young and gather the edible plants in the vicinity. The food isn't consumed as it is found, but is brought back to camp to be shared by everyone. The early hunter-gatherers were undoubtedly nomadic, changing camps as the local vegetation and game were exhausted.

The transition to the highly efficient hunting-gathering economy must have been slow and painful, since it requires a high level of social control over an individual's impulses. There is nothing instinctive about it. Desertion of wife and children may be a phenomenon of modern times, but a *H. erectus* male must have had the same impulse to skip out when the going got tough. Only groups that developed strong social control could keep the male hunters bringing home the bison.

As females became more dependent on males for food for themselves and their children, female competition for males as permanent providers would have increased. A female who was sexually available at all times would have a decided advantage in holding a male provider over one who was receptive for only a few days each month. Natural selection, operating within the culturally determined hunting-gathering environment, would then favor females whose sexual receptivity was independent of their estrous cycle. In this scenario, human sexuality is seen as arising in response to a cultural invention, the hunting-gathering economy.

Other human characteristics may have evolved in response to the same selective pressure. For example, the female breasts, which mimic the buttocks, could have been a sexual attractant to front-mounting males still imbued with the instincts of their rear-mounting ancestors. Hairlessness may have also evolved in females because they were more attractive naked. Many peculiar animal characteristics—from the peacock's tail to the proboscis monkey's nose—evolved by means of sexual selection, so it's plausible that hairlessness evolved by the same mechanism. And plausible is all it is, because there is absolutely no evidence as to what the actual selective pressures acting on our ancestors were. Some were certainly environmental. But some must also have come from the *Homo*-made culture itself, so culture would have had a very strong influence on *Homo*'s subsequent genetic evolution, clearly a case of a species making itself.

Sometime within the last million years, *H. erectus* became the first hominid to leave Africa. Fossil remains dating from 250,000 to 750,000 years ago have been found at sites in Indonesia ("Java Man"), China ("Peking Man"), and Europe. The rapid expansion of a heretofore relatively localized species indicates either a sudden change in cultural habits or climate or both. Perhaps there was a spread of grasslands into northern Africa, southern Europe, and Central Asia, which the hunter-gatherer groups could follow in their reach for game. An advance of only ten miles per century would bring *H. erectus* from Tanzania to Java in 70,000 years. Perhaps an occasional group developed wanderlust and just started trekking over existing grassland. Such long-distance colonization could have established *H.*

erectus in China in one generation, since a person can walk from Africa to China in a few years. In any event, a million years elapsed before modern humans, another large-brain species, spread across the Old World.

Homo sapiens

There is now strong evidence, both molecular and fossil, that our own species, *Homo sapiens,* evolved in Africa about 200,000 years ago (Gibbons, 1992b; Lewin, 1987a, 1987b; Stringer and Andrews, 1988).[2] Genetic studies of existing human populations show that humans worldwide have much less genetic variation than is found in a single gorilla subspecies, indicating that *H. sapiens* evolved very recently from a small isolated population, perhaps from a single individual.

Although most DNA is in the chromosomes of a cell, a small amount is also present in the mitochondria, small subcellular units outside the cell's nucleus. During fertilization, only the sperm's chromosomal DNA enters the egg; all the mitochondria of the fertilized egg, and of the individual that grows from it, come from the mother. It has been found that all human beings have nearly identical DNA in their mitochondria, indicating descent from a common female about 200,000 years ago. What small variation there is, is greatest among the peoples of sub-Saharan Africa. This indicates that modern humans probably originated in Africa, since the amount of variation in a population's mitochondrial DNA increases with time.

A new technique, called thermoluminescence, has been used to date burnt flints found in the 1930s with human skulls in a cave in Qafzeh, Israel. Over time, radioactive decay in the surrounding rock produces tiny changes in the atomic structure of the flint. The older the material, the greater the changes. But when the flint was burnt, presumably by the people whose bones were found with them, these changes were erased. When a piece of burnt flint is heated in the laboratory, light is emitted in proportion to the amount of changes produced since the flint was burned. In this way, the flints associated with the Qafzeh skulls were found to be 92,000 years old, twice the age of any previous human material (Valladas et al., 1991).

The picture that is emerging from these studies is that *H. sapiens* evolved in Africa 200,000 years ago, presumably from an isolated *H. erectus* population. Over the next 100,000 years, the new species spread throughout Africa, reaching North Africa about 100,000 years ago. From there, the population spread into western Asia (Iran and Iraq), reaching Europe about 40,000 years ago. There it met another human population, the Neanderthals.

Until recently, the well-studied Neanderthals were thought to be direct ancestors of modern humans, perhaps a subspecies of *H. sapiens*. This is now in doubt. The Neanderthals probably evolved independently from *H. erectus* and thus are only cousins of ours. They had the 1.4-liter brain capacity of modern humans, but the low forehead and prominent browridge of *H. erectus* and an even more protruding muzzle. Their frame was robust, with heavy bones and large muscles. The recent discovery of a complete Neanderthal pelvis from the Kebara cave in Israel shows that their pelvic outlet was shaped differently from that of humans and that the orientation of the sockets into which the thigh bones fit was different as well. This latter may well indicate that the Neanderthals stood and walked quite differently from modern humans.

Nevertheless, the Neanderthals were undoubtedly human. They had a stone industry, called the Mousterian, that was much more advanced than the Acheulian. Instead of shaping a rock by striking flakes from it, the tools were made from large flakes carefully struck from a core rock. In this way, many extremely delicate blades and knives could be made from one rock, increasing the efficiency of material use by a factor of fifty or more. Furthermore, the Neanderthals built shelters that allowed them to survive the frigid winters of Central Asia, and they buried their dead with food and tools, indicating some spirituality.

The Neanderthals and their culture disappeared within 5,000 years of the arrival of modern humans into Europe and Asia. Modern humans are considerably less robust than the Neanderthals, with smaller bones and muscles, a high forehead, a flattened muzzle, and no browridge. How, or whether, this newer variant was superior either biologically or culturally to the Neanderthals is not known. Modern humans' early European cultures, the Perigordian and the Aurignacian, show an increase in skill and technique with stone, but no major innovation.

But, just as their ancestors 3 million years earlier had learned to live in small groups with several adult males, the moderns learned to organize these groups into tribes. At times, the men from different groups would band together for ceremonies and war. Language, perhaps originating as nursery talk among women and children, could have been the critical factor in uniting a tribe and forging the first primitive armies. Isolated Neanderthal groups, with three or four languageless men, would have been no match for an organized attack by a dozen or more moderns able to shout orders to one another.

There is nothing in this description that assumes the moderns were superior to the Neanderthals. It merely assumes they had learned a few skills that gave them a military advantage. The advantage need not have been large. They did not have to fight often or win every battle. If the moderns replaced only 2 percent of the Neanderthals in each generation, the Neanderthals would be virtually extinct in 5,000 years.

Disease is another possible explanation for the moderns' success. The moderns could have carried a disease to which they were immune, but to which the Neanderthals were susceptible. Even a small relative difference in susceptibility could, over a few dozen generations, give the moderns a decided population advantage over the more susceptible Neanderthals. Or disease could merely have weakened the Neanderthals to the point that the moderns could defeat them more easily in battle. Unfortunately, the fossil record is too scanty either to support or to reject disease as a critical factor in the success of the moderns over the Neanderthals.

We do know that tribal migrations and conquests have been universal among humans for at least the last 5,000 years. Therefore, it is a very small extrapolation, given the time scales we have been considering, to assume that tribal warfare existed 40,000 years ago. Furthermore, from the example of the Zulu chieftain Shaka Zulu (1787–1828), we know that a single man can teach a small tribe to fight so much more effectively than its neighbors that in only a few years, the depopulation of a vast area results. Because of the depopulation of the Natal that resulted from Shaka's wars of extermination, the Boers were unopposed on their Great Trek through that territory in the 1830s.

That warfare should be such a characteristic of human beings seems inconsistent with the idea that the *Homo* line had developed the ability to control male rivalry. Isn't war the exercise of male rivalry par excellence? Indeed! But war is possible only because the warriors in each army have learned to control their intragroup aggression and to direct it at the males of another group. The irony is that men must cooperate in order to achieve the destructive power of organized war. That is why we see in war the best and the worst that is in us and why we honor our soldiers while we pray for peace.

Over the past 40,000 years, humans have extended the prohibition against intragroup violence to ever-larger groups, enabling people to live in villages, towns, and, eventually, large cities. As the size of these units increased, more formal methods of control were needed, culminating in our complex criminal justice system. Street gangs and urban violence attest to how close to the surface the primitive forms of male aggression are. Most men have learned, however, to redirect male rivalry from physical aggression to economic or athletic aggression. This redirection is one of the major accomplishments of our species. Organized society, far from having robbed human beings of some original purity or innocence, has in fact given us whatever claims we may have to nobility. But the ability to cooperate in building complex cities is also the ability to cooperate in building armies, and the latter undoubtedly preceded the former.

There appears to have been an almost explosive expansion of modern humans 40,000 years ago. Not only did they displace the Neanderthals from Europe and Asia within a few thousand years, but they also reached Australia at least 33,000 years ago. There is incontrovertible evidence that humans reached America 13,000 years ago, but the evidence for an earlier arrival is in doubt (Dillehay, 1991). Perhaps some humans reached the Americas 30,000 years ago, but didn't survive long enough to leave indisputable evidence.

The reason usually given for these huge migrations is that the hunter-gatherer groups were following the herd animals across the expanding grasslands. But if food had been the primary motivation to migrate, there would have had to have been intense population pressures at home. There is no evidence of this. More likely, the early moderns, like ourselves, sometimes got the urge to pick up and travel thousands of miles to start a new life. After all, we are a nomadic species, always on the move for greener pastures. The behavior of explorers, emigrants, and colonists attests to our nomadic spirit. From what we know of the early moderns, peoples in historic times, and ourselves today, migration has always been an intrinsic part of human behavior.

Historical analyses often undervalue the significance of human mobility. Ideas and innovations that originate in one place may well diffuse slowly from village to village, but they can just as well be carried by a single person, a Marco Polo, across thousands of miles in a few years. Trade goods have been found in some of the earliest sites of modern humans, indicating that 40,000 years ago there were traveling salesmen who journeyed great distances carrying goods, ideas, and genes. Isolated human settlements may appear stable when viewed on a time scale of a few generations, but when looked at on a time scale of hundreds of years, most have come from somewhere else or been visited or invaded frequently by strangers.

The Neolithic Agricultural Revolution

Near the end of the Ice Age, about 18,000 years ago, hunting conditions were excellent in southern Europe. Herds of reindeer and bison supplied ample food to the skilled hunters. From this period come the exquisite cave paintings of Altamira in Spain and Lascaux in southwestern France. Bison and other animals are the main subjects of these paintings, often very realistic in their depiction, whereas humans are infrequently drawn, and then only as stick figures. Whether they were used for hunting magic or for clan rituals, these paintings indicate a rich cultural life. In fact, despite the cold climate, the Upper Paleolithic was a high point in human history that wasn't to be surpassed until the time of Egypt.

The hunting-gathering economy is a surprisingly efficient one. Even today, in the inhospitable Kalahari desert, the nomadic !Kung work only a few hours a day securing food, with a minimum set of tools and implements. With the abundant game of the Upper Paleolithic, men would have had even more leisure, and work, as we know it, would have been unknown.

As the climate warmed, the herds diminished, and most of the succeeding Mesolithic cultures maintained a hunting-gathering economy based on smaller and less glamorous prey. To preserve their access to a dwindling meat supply, some groups started to protect small bison herds from predators and other hunters. Following the herds at first, these groups eventually learned to lead them between winter and summer feeding grounds. For these groups, the domestication of cattle replaced a nomadic hunting-gathering economy with a more secure nomadic hunting-gathering-herding economy. This brilliant adaptation to a changing environment, though it did require the social discipline and forethought to eat only as much of the herd as could be regenerated each year, probably required little additional work.

Then, suddenly, about 10,000 years ago, at a few isolated sites in the Middle East, the Neolithic agricultural revolution erupted. Some of the oldest archaeological evidence of this comes from Jericho, a site in the Israeli-occupied West Bank territory. Situated 825 feet below sea level and only six miles from the Dead Sea, it seems the least likely spot on earth in which to invent farming. But in the first century A.D., the Jewish historian Josephus described the abundant spring at Jericho as "admirably suited for irrigation" (90/1970, p. 271). In 9000 B.C., it was just a campsite for Mesolithic hunter-gatherer-herders, who were attracted by the spring. By 8000 B.C., it was a walled town covering seven acres (three hectares) and inhabited by several thousand people. Grains of cultivated wheat and barley have been found at this site, making it one of the oldest known agricultural communities in the world.

What could have caused such an abrupt change? Arguments about population pressure or climate changes don't suffice, because the agricultural revolution was for a long time limited to the Fertile Crescent, a small geographic region that runs from Egypt through Palestine to Iran and Iraq. This is an arid region that has always required irrigation to make fertile, and the population of the region didn't increase appreciably until after agriculture was well established.

Again, it is probably in the human social environment, rather than the physical environment, where we will find a stronger explanation. As solitary hunter-gatherer groups banded together into tribes, an increasingly complex hierarchical social organization developed. At first the chieftains, shamans, elders, and warriors exercised authority only in times of war, which were probably rare during the Upper Paleolithic. But as herding began to replace hunting and human beings for the first time had something besides women to steal, raids and wars became more frequent. After all, to a hunter from a nonherding tribe, a herdsman's cow is just another bison. The domestication of cattle must have increased warfare, and, in fact, scenes of warfare appear for the first time in Mesolithic art.

In such an environment, the need for continuous defense gave permanent power to the warrior class and its attendants. This, in turn, led to the hierarchical political system, with the commander in chief of the military forces as the leader, that has been universal ever since. A prosperous herding community needed a strong military force that could defend its pastureland and mount annihilation raids against encroaching nomads. The ancient civilizations of Egypt, Rome, and China were at constant war with the tribesmen on their borders. And so, too, must have been the earliest Mesolithic herdsmen.

The more prosperous a community became, the greater were its defense needs. At some critical point, the need for defensive structures exceeded the community's ability or willingness to build them. Remember, work as we know it didn't exist for the men of a hunting-gathering-herding community. (Wherever modern civilization has encroached on indigenous hunter-gatherers in Australia, Africa, and North America, major social problems have arisen from the difficulty the men have in adopting our notions of work; women, of course, work in all communities.) To solve their labor problem, therefore, instead of killing the men they defeated in battle, the herding community's warriors started putting them to work building their defensive fortifications. Although we think of slavery as evil, at the start it was a substitute for killing captives. With the invention of slavery, a male captive came to have as much value as a woman or a cow.

Since the slaves built the defensive walls, the more slaves a community obtained, the more security it would have. Soon the slave population would have grown beyond what could be supported by gathering the wild grains that grew around the local river or spring. Canals were then dug to irrigate arid land away from the river, enlarging the area that could grow grain. And thus the Neolithic agricultural revolution was initiated. The ruling class undoubtedly enjoyed bread with their meat, but the main purpose of irrigated grain agriculture was to supply food for the slaves. It was a natural step from domesticating animals to domesticating people.

In Jericho, irrigation canals brought water from the spring to the arid fields. Labor and energy of unprecedented amounts would have been needed to dig these canals and to build the half-mile-long stone wall and watch tower that guarded this very early city-state. Solid houses replaced the temporary dwellings of the Mesolithic period, attesting to still more intensive labor. The only economic system of the period that could have organized this much labor is slavery.

Large-scale slavery requires a hierarchical social organization of considerable sophistication. There were no chains in Neolithic times, so the captives would have been bound by custom and law, perhaps by a state religion that created fear and dependence, and certainly by protection from invading nomads who might not value a live slave as much as did their civilized overlords. After a few generations, the slaves would have lost the skills needed for their former life-style and would be as dependent as their masters were on the slave economy.

Wheat and barley cultivation started in the Middle East 10,000 years ago; rice and millet cultivation started in China 7,000 years ago; and maize (corn) cultivation started in Mesoamerica 5,000 years ago. In all cases, the staple crop was the seed of a member of the grass family. Grass, which had driven our ape ancestors from the trees and had nourished the large herds that our Paleolithic ancestors hunted, was the agricultural staple of our Neolithic ancestors. And grass, whether eaten directly as grain or sugar (from sugarcane, a grass) or indirectly from grass or grain-fed animals, is still a major portion of the human food supply.

The Neolithic agricultural revolution was one of the most important episodes in human history, absolutely essential to the development of civilization. It's wrong, however, to think that it was an advance on a previous economy. Agriculture was needed to support the pockets of dense population (towns) that arose to defend against raiding nomads and rival tribes. Much of this population would have been slaves or refugees forced by the exigency of war to leave their Mesolithic life for a Neolithic one. Like the descent of our ape ancestors from the trees, agriculture arose from grim necessity. After all, the hunting-gathering-herding economy it replaced is a relatively free and easy one. There is a lot of leisure; the work isn't too hard; and meat is eaten frequently. As a Tibetan nomad, living today in one of the most barren and ferocious places on earth, explained to anthropologists: "Look, it is obvious that we have a very easy life. The grass grows by itself, the animals reproduce by themselves, they give milk and meat without our doing anything. So how can you say our way of life is hard?" (quoted in Goldstein and Beall, 1989, pp. 764–66). By contrast, farming is hard work, with little leisure and a diet mostly of gruel (Diamond, 1987). Human beings didn't switch from one to the other of their own free will.

Up to this point, nothing in the narrative of human evolution and history foreshadowed the great technological triumphs that were to come 10,000 years later. Toolmaking doesn't necessarily lead to agriculture; agriculture doesn't necessarily lead to large cities; and large cities don't necessarily lead to science. *Homo erectus* survived for nearly a million years in the Stone Age, and humans survived for 200,000 years without agriculture.

The twists and turns on the journey from apes to agriculture were caused by myriad independent local factors, most of which we can never know. The journey from early civilization to science was no less tortuous. This is made clear in Chapter 4, which examines the earliest writings of Greece and Israel. These writings reveal the cultural factors that caused these neighboring civilizations to develop in radically different directions.

4
Prophets and Poets

Cultural Limits

Every culture develops its own words, concepts, and mental operations out of its own peculiar history and circumstances. Once formed, they limit the type of activities and institutions that the culture can create. Therefore, the failure of a culture to develop a certain activity may not be because it didn't think of it but because it couldn't think of it.

The individuals in a culture are stuck with the mental structures of that culture. New structures can't be acquired from within the culture because there will be no activities from which to learn them. Counting can't develop in a society that has nothing to count, nor can any ability for which counting is a prerequisite.

Most of the higher-level activities of advanced cultures require a complex of structures involving models of reality, social values, and specialized skills. Debating, for example, requires the acknowledgment of rational minds besides one's own, the acceptance of discourse as an alternative to force, and the attainment of the rudiments of persuasive argumentation. These and a dozen other factors are necessarily a part of any culture that engages in debate, and they in turn are likely to have led to the development of still higher structures, such as logic, political philosophy, and scientific speculation.

All this is to say that a higher-level cultural activity can't exist in isolation. The Piagetian view that the mind is organized into a hierarchy of mental structures implies that a higher-level activity is accompanied by a host of associated elements. Conversely, the absence of most of the necessary elements of an activity indicates not only that the activity doesn't exist at all in the culture, but also that the mental structures required for the activity don't exist as well.

In this view, science, which is a very high-level activity, required a set of prerequisite structures. What were they? What activities were associated with them? Why did some cultures develop them and not others? Was Greece the only culture ever to have developed these structures? And if so, why? These are the questions we wish to address in this chapter. Although they are very deep, we have a powerful premise with which to attack them: The prerequisites for science either permeate a culture or are absent altogether.

Greece and Israel

We shall concentrate on the two cultures that form the foundation of Western civilization, those of Greece and Israel. Although no two civilizations could be more dissimilar, both based their earliest myths on events that occurred on the eastern shore of the Mediterranean Sea sometime during the thirteenth century B.C. One was the siege and destruction of Troy, a fortified city on the northwest coast of Anatolia, by an armada of Mycenaean warriors. The other was the conquest of Canaan by Hebrew-speaking tribes migrating eastward from the Nile valley.

The destruction of Troy itself had no lasting political consequences of which we are aware. We know of it only from the legends that have come down through the ages, especially the Homeric epics. Written some 500 years after the fall of Troy, the *Iliad* and the *Odyssey* are the world's oldest continuously read works of fiction and, in many respects, still the greatest. Homer was the principal source of classic Greek culture, providing the language, religion, and attitude that was to characterize Hellenic civilization for 1,000 years.

Similarly, the Israelite conquest of Canaan had no influence on the great powers of the time. Israel, even at its height under Solomon, was hardly noticed by the ancient chroniclers. Yet it too produced a book, the Old Testament, that, through Christianity, became for Europe what Homer had been for ancient Greece.

After the conquest of Palestine by Alexander the Great in 332 B.C., Israel came under Greek control and influence. Many Jews were attracted to Greek rationality, and Jewish communities were established in Greek cities throughout the Middle East. A particularly influential Jewish community existed in Alexandria, the center of Greek science and learning in the last three centuries before the Christian era. There, in the third century B.C., Jewish scholars translated the important books of the Jews into Greek. This collection, known as the Septuagint, is the Greek version of the Old Testament. The official Hebrew version wasn't compiled until about A.D. 100. Both versions were based on much older material and so give us good insight into the thinking of the ancient Israelites. However, even the oldest books of the Old Testament may not be entirely free of Greek influence.

Homer and the Bible give us ready access to the roots of the Hellenic and the Judeo-Christian cultures. The Bible exemplifies egocentric thinking at its purest. Self-appointed prophets claim personal knowledge of God's will, which they exhort the people to obey. The claim is never questioned, even rhetorically. Such claims of personal knowledge are very much with us today, from papal encyclicals to television evangelicals.

Homer is the world's first exemplar of objective thinking. In the *Iliad*, the assembly, in which men of different rank debate matters of mutual concern, is a hallowed institution, and skill in debate is held as high as skill in war. The gods try to control human affairs, but they are often at cross purposes, so their efforts cancel out one another. In the end, human affairs are worked out by humans according to their own precepts of right and wrong.

Ancient Israel

About 1372 B.C., the heretic Pharaoh Amenhotep IV succeeded his father to the throne of an imperial Egypt whose territories included Syria and Nubia. Perhaps projecting onto the supernatural the magnitude of his own unparalleled imperial power, Amenhotep recognized only one god, the sun god Aton, and believed himself to be his son. He changed his name from Amenhotep ("Amon is satisfied") to Akhenaten ("Aton is benefited") and, with the intolerance that ever after has characterized monotheism, suppressed the worship of the traditional Egyptian gods, especially the popular Amon.

After his death in 1354 B.C., the Aton cult was abolished and the traditional polytheistic Egyptian religion reestablished (Weigall, 1910).[1]

Akhenaten's radical theological departure had important consequences on the art of his time, which became much more lively and naturalistic. The bust of his wife Nefertiti, now in the State Museums in Berlin, is one of the most hauntingly beautiful portraits of all times. Whether his theology influenced the subsequent development of monotheism is less certain. It may have influenced the resident Hebrews, whose exodus from Egypt is thought to have occurred a century later (Freud, 1939/1955). More likely, both Akhenaten and Moses were influenced by a penchant for monotheism or, rather, monolatry, among the many small nomadic tribes scattered throughout the surrounding desert.

Monolatry is the worship by a tribe of a particular god who has taken the tribe under his protection. The existence of other gods is taken for granted. Each alien tribe is assumed to be under the protection of its own local deity, just as each is under the leadership of its own patriarch. Monolatry is a projection of the relatively simple patriarchal tribal system onto the supernatural and as such is less sophisticated than the elaborate polytheistic religions of the literate urban civilizations. Monolatry may have been the form of religion of many of the small tribes and kingdoms—Ammon, Philistia, Aram, Moab, Edom, and the tribes of Israel—that occupied the Levant and the Jordan valley around 1000 B.C. At that time, Yahweh was just the god of Israel, as Dagon was just the god of Philistia. He was, perhaps, a bit touchier than the others: "Do not follow other gods, gods of the people around you, for Yahweh your god is a jealous god; the wrath of Yahweh your god would blaze out against you, and he would wipe you off the face of the earth" (Deut. 6:15).

The Israelites, like their neighbors, practiced a narrow tribal religion. However, the extreme jealousy of Yahweh caused the Israelites to become increasingly insular:

> When Yahweh your god has brought you into the country which you are going to make your own [Canaan], many nations will fall before you. . . . You must put them under the curse of destruction. You must not intermarry with them; you must not give a daughter of yours to a son of theirs, or take a daughter of theirs for a son of yours, for your son would be seduced from following me into serving other gods; the wrath of Yahweh would blaze out against you and he would instantly destroy you. (Deut. 7:1–4)

Perhaps more than other people, the Israelites tried to regulate their civil life according to religious law. This Law was given to Moses by Yahweh, as recorded in Deuteronomy 1:1–5:

> These are the words which Moses addressed to all Israel beyond the Jordan, in the desert, in the Arabah facing Suph. . . . It was the fortieth year, on the first day of the eleventh month, that Moses told the Israelites everything that Yahweh had ordered him to tell them.
>
> . . . There, in Moab beyond the Jordan, Moses resolved to expound this Law.

Originally an oral tradition, Deuteronomy was first written many centuries after Moses and edited and reedited many times thereafter. In the form in which it has come down to us, it is the Law for a settled pastoral people, not for the band of invading nomads that Moses was leading. Its prohibition against intermarriage probably dates from after the exile of the Jews to Babylonia (586 B.C.), some 700 years after Moses. For in the days of the judges, no such proscription prevented Samson from marrying a Philistine or Boaz from marrying the Moabite Ruth, the great-grandmother of David.

In Deuteronomy 18:18–20, Yahweh does provide a mechanism for changing the Law:

> I shall raise up a prophet like yourself [Moses]. I shall put my words into his mouth and he will tell them [the Israelites] everything I command him. Anyone who refuses to listen to my words, spoken by him in my name, will have to render an account to me. But the prophet who presumes to say something in my name which I have not commanded him to say, or who speaks in the name of other gods, that prophet must die.

Judaism is the oldest of the historical–prophetic religions, and this passage is the earliest authorization of the prophets as the spokesmen of God. The later historical–prophetic religions—Christianity and Islam—derive their prophetic tradition from Judaism and share with Judaism the basic theory of knowledge that is expressed in the Old Testament.

The theory is simple: All knowledge comes from God and is revealed to humans through prophets. In preexilic times, the prophets give specific instructions regarding affairs of state:

> Samuel said to Saul, . . . "This is what Yahweh Sabaoth says, 'I intend to punish what Amalek did to Israel. . . . Now go

> and crush Amalek; put him under the curse of destruction with
> all that he possesses. Do not spare him, but kill man and
> woman, babe and suckling, ox and sheep, camel and don-
> key.' "(1 Sam. 15:1–3)

Later, with the calamities of defeat and exile that afflicted Israel, the
prophets became more abstract, more moral, more poetic:

> For Jerusalem has collapsed and Judah has fallen, because their
> words and deeds affront Yahweh and insult his glorious gaze.
> Their complacency bears witness against them, they parade
> their sin like Sodom, . . .
> . . . O my people, your rulers mislead you and efface the
> paths you ought to follow! Yahweh has risen to accuse, is
> standing to pass judgement on the people. Yahweh is about to
> try the elders and the prince of his people. . . . "By what
> right do you crush my people and grind the faces of the poor?"
> says the Lord Yahweh Sabaoth. (Isa. 3:8–15)

The postexilic prophets also tended to be tedious and diffuse.
They failed utterly to create a new view of Israel's place in a changing world, but fell back on the millennia-old notion that everything
bad that happened was Yahweh's punishment for Israel's wicked-
ness.

The prophetic voice, from Moses to Muhammad, has been an
emphatically injunctive one. The prophet doesn't attempt to persuade by reasonable arguments, but merely proclaims the revealed
truth:

> Praise be to Allah who has revealed the Book to His servant
> shorn of falsehood and unswerving from the truth, so that he
> may give warning of a dire scourge from Him, proclaim to the
> faithful who do good works that a rich and everlasting reward
> awaits them, and admonish those who say that Allah has be-
> gotten a son. (Koran 18:1–5)

The prophetic voice is so egocentric that it doesn't acknowledge the
existence of other reasoning minds. In such a cultural setting, rea-
soned discourse is meaningless. The Koran and the Bible (before the
Greek period) lack even the words normally associated with debate:
reason, credible, impossible.

There is one interesting exception. In Deuteronomy 18:21–22,
after Yahweh tells Moses he will send other prophets in the future,
he comments:

> You may be privately wondering, "How are we to tell that a
> prophecy does not come from Yahweh?" When a prophet
> speaks in the name of Yahweh and the thing does not happen
> and the word is not fulfilled, then it has not been said by Yah-
> weh. The prophet has spoken presumptuously. You have noth-
> ing to fear from him.

This acknowledgment of private thoughts—of skepticism—is strikingly out of place with the injunctive tone of most of the Old Testament. The use of a rhetorical question is a debater's device of some
sophistication. The attempt to establish a reasonable basis for distin-
guishing a true prophet from a false one shows an appreciation of
the dilemma inherent in the prophetic tradition. Deuteronomy is
known to be a late book, and it is possible that this passage is the
result of Greek influence.

With the destruction of Jerusalem and the exile of its elite to
Babylon in the sixth century B.C., the religious and prophetic institutions of Israel were destroyed. To preserve their culture, religious
scholars began to collect and edit the oral and written traditions of
their people. About this tradition, we know today only those edited
collections that were eventually included in either the Septuagint,
the third-century B.C. Greek translation, or the first-century A.D.
official Hebrew compilation.

By the time of Jesus, Palestine had been under Greek influence
for over 300 years. So although the New Testament is written in
the old prophetic style, the Greek influence can be found in many
places. Jesus, who is generally equivocal in his remarks, loses many
of his followers when he makes an uncharacteristic attempt at mysticism:

> As the living Father sent me and I draw life from the father, so
> whoever eats me will also draw life from me. This is the bread
> which has come down from heaven; it is not like the bread our
> ancestors ate: they are dead, but anyone who eats this bread
> will live for ever. (John 6:57–58)

"This is intolerable language. How could anyone accept it?" his followers declared. Apparently, after 300 years of Greek rule, the Jews
had developed a certain amount of skepticism. Jesus seemed to learn
from this incident, for thereafter he kept to the main mission of any
prophet: to be accepted as a true emissary of God.

Judaism, Christianity, and Islam are monotheistic religions based
on the teachings of prophets. They share the belief that some human

beings can directly know the mind of God, but they disagree on which of them are the true prophets and which are the false ones. Since there is no rational way to decide this, conflict among prophetically based religions is inevitable, thereby driving monotheism further into irrationality and intolerance. Open inquiry, along with alternative approaches to the supernatural, are forbidden, as they would undermine the power of the religious authorities.

Western religious thinking is based directly on Eastern mysticism, all claims of Western rationality notwithstanding. If anything, the Christian European civilization that arose among illiterate Gallic and Germanic tribes was less advanced than its Eastern counterpart, which had been literate for 1,000 years before Caesar conquered Gaul. Today mysticism is still a strong component of our culture, openly accepted by religious and spiritualists of all faiths.

Scientific thinking didn't—and couldn't—evolve from the prophetic tradition of Judaism and Christianity; it arose from a totally different tradition.

Mycenaean Civilization

In the *Iliad,* Agamemnon, leader of the Greek expedition to Troy, is said to be the king of Mycenae, a city rich in gold. There is an unimportant town by that name in southern Greece, and it was here, in 1876, that a German businessman-turned-archaeologist, Heinrich Schliemann, unearthed a circle of grave shafts containing immense quantities of exquisitely crafted gold objects. Further excavation there and at other sites revealed that around 1400 B.C. a highly developed Late Bronze Age civilization flourished in Greece and on the island of Crete. Now called the Mycenaean civilization, after the site in Mycenae, it consisted of a number of rich and powerful kingdoms that shared a language and script. Clay tablets written in the Mycenaean script, called Linear B, were first uncovered in 1900, but because neither the script nor the language was known, they remained undeciphered for fifty years. All that the scholars of the time could agree on was that the language wasn't Greek.

Michael Ventris, a young English architect, began serious work on Linear B in 1950, believing that the language might be related to Etruscan, a still-undeciphered ancient Italian language. Nevertheless, he proceeded by using an ingenious analysis of word endings to eke out the phonetic values of a few symbols. From the context of the tablets, the words for *boy, girl,* and *total* (sum) could be recognized, and when he tried his values in them, they sounded like an archaic form of Greek. He soon joined forces with the Greek philologist John Chadwick, and together they completed the decipherment of Linear B, proving that the highly developed second-millennium Mycenaean civilization spoke Greek.[2] Documents from this period were found to contain the names of some of the gods of the classical period: Zeus, his wife Here, Poseidon, Hermes, Athena, and Artemis. Thus some elements of the later Greek culture were in place 700 years before Homer (Chadwick, 1968).

These Mycenaean kingdoms flourished from 1400 to 1200 B.C. and then went into decline, vanishing completely by 1100 B.C. Little is known of the subsequent three centuries, except that it was a period of transition, during which Dorian invaders became the ruling elite. By 800 B.C., small independent city-states emerged, with economies based on maritime trade and plunder. This period set the stage for the next 1,000 years of Greek civilization, because from it came the *Iliad* and the *Odyssey*.

Homer

Working with traditional legends and stories from Mycenaean times, Homer created masterpieces of original fiction. Their mere existence presupposes a prosperous culture that supported professional bards and encouraged the creation of new works, just as the existence of Shakespeare's plays required a vibrant theater in Elizabethan times. Carrying the analogy further, we may assume that the audience for Homer's works wasn't just an elite of aristocratic warriors, but a large class of wealthy merchants and traders as well. For even though Homer exalts the warrior ideals of valor and martial skill, the happiness of family life is held in greater esteem.

Great efforts have been made to determine whether the major events in the *Iliad* actually took place (Wood, 1985). Seeking the ruins of Troy itself, Schliemann discovered the ruins of a second-millennium city in northwest Turkey several years before his discovery at Mycenae. But although Schliemann's work proves that Mycenae was a great kingdom around 1250 B.C., it doesn't prove that Agamemnon lived there or that he led a thousand black ships from Greece on a ten-year siege of the city that once stood on Schliemann's site in Turkey. Nor does it matter much to our study.

What does matter is that the archaeological evidence is fully consistent with Homer. So whether the old legends were about a particular war—highly likely—or were about the wars in general that the Greeks waged along the coast of the eastern Mediterranean, Homer's account is about a real period in history.

Nothing is known about Homer himself. Before the decipherment of Linear B, most scholars thought that the Mycenaeans didn't speak Greek. It was therefore difficult to understand how, out of foreign materials, a Greek masterpiece could have been produced at the very beginning of Greek civilization. Homerian scholars saw incongruities in the *Iliad* and the *Odyssey* that suggested to them that neither was the work of a single person. Some claimed that the works grew by accretion, as succeeding generations of bards added new material; others believed that Homer was merely a compiler who pieced together a body of traditional legends and song cycles. However, there was no agreement on this or any objective evidence. By the 1930s there were more opinions on the matter than there were Homerian scholars, as many scholars had more than one opinion (Nilsson, 1933). Now that we know that a Greek civilization existed for seven centuries before Homer, the emergence of a poet of his magnitude is no longer a problem. Over a century of literary analysis failed to prove that the *Iliad* and the *Odyssey* are other than what they appear to be: masterpieces by a great poet.

Scholars will never be able to prove otherwise, because the *Iliad* and the *Odyssey* are far more consistent than they needed to be. They are epic poems that were first sung by professional bards at private parties and public ceremonies. No more than a chapter or two could have been recited on any one occasion, and the audience wasn't expected to remain sober throughout the performance. The many wine-drinking scenes in the *Iliad* and the *Odyssey* were probably cues for the listeners to refill their cups. Moreover, few would have ever heard the entire work in sequence.

Yet in spite of this, the *Iliad* and the *Odyssey* are unified works, each with a beginning, a middle, and an end. The oldest works of fiction in Western literature, they appear fully formed: characters with individualized personalities, plots and subplots that move forward inexorably, and scenes that shift from the Achaeans to the Trojans to the Olympians, with corresponding changes of mood and setting. For example, the chariots driven by humans have axles made of wood, whereas the chariots of the gods have axles made of iron, the rare new wonder metal of the Late Bronze Age.

Contrast this with the Old Testament, a work we know is a compilation of hundreds of stories written over many centuries. It abounds in inconsistencies and repeated passages that don't require a scholar to discover. Consider the familiar story of David and Goliath. As a boy, David is anointed king by the prophet Samuel, but then becomes armor-bearer to Saul, the current king (1 Sam. 16). A few verses later (1 Sam. 17:12–30), David is again just a shepherd, bringing food to his brothers at the front, when he hears of Goliath's challenge. He is back in Saul's service in the following verses (1 Sam. 17:32–39) and is again unknown to Saul after killing Goliath (1 Sam. 17:55–56). Logical consistency, or noncontradiction, didn't seem to be of concern to the Hebrew editors.[3]

Noncontradiction is, however, of major concern to Homer. His epics aren't just stories of great deeds, but are a complete cosmography. From the division of the world among the three sons of Cronos to the preference of young men for the latest songs, every matter of concern to man or god is considered without mysticism or equivocation. When the ill-tempered Ares, the god of war, learns that his human son Ascalaphus has been killed in battle, he flies into a blind rage. Athena councils him: "Take my advice then, and set aside all thoughts of avenging your son. Many a finer and stronger man than he has been killed before now and will be killed hereafter. We [gods] can hardly be expected to keep our eyes on the pedigree of every man on earth" (*Iliad* xv, p. 275). Thus does Homer explain, at the very beginning of Western civilization, how evil can befall a good man.

The *Iliad* tells of events in the ninth year of the Trojan War, about a year before the city was destroyed. The *Odyssey* tells of the adventures of Odysseus (Ulysses in Latin) as he makes his way home after the war. One is a tragedy about the futility of war, the other a sea yarn. The somber *Iliad* is told to the resonating beat of the kettledrum, whereas the fanciful *Odyssey* is sung to the lyric notes of the flute. As the translator E. V. Rieu noted, a reader "will be brought closer to tears by the death of a single horse in the *Iliad* than by the killing of the whole gang of Suitors" in the *Odyssey* (*Iliad*, pp. vii–viii). One can well imagine that the *Iliad* was composed in a time of war and chaos, so vainly do its heroes fight and die on the Trojan plain. The *Odyssey*, I like to think, was composed in Homer's later years, a time of peace and stability.

It is clear from the *Iliad* that by Homer's time the Greek custom of open debate in assemblies was well established. The opening scene of the *Iliad* is, in fact, an assembly attended by all the Achaean warriors, nobles and commoners alike, to decide what to do about a deadly plague that has afflicted them. The wisest seer in the camp is just a common soldier, who explains that the plague is a punishment from Apollo for Agamemnon's refusal to accept a ransom in exchange for the captive girl Chryseis, the daughter of Chryses, a priest of Apollo. A heated argument ensues between Agamemnon and Achilles over the compensation that Agamemnon should receive for returning his beloved Chryseis to her father. When none is agreed on, Agamemnon declares that he will take Achilles' prize, the beautiful Briseis. At this threat, Achilles reaches for his sword but is restrained by the goddess Athena, who admonishes him: "Come now, put up your strife and take your hand from your sword. Sting him with words instead, and tell him what you mean to do. . . . Hold your hand, then, and be advised by me" (*Iliad* i, p. 28).

Homer's highest praise for a man (besides having comely feet) is that he is good at war and debate. In the assembly, even the lowly Thersites ("the ugliest man that had come to Ilium") could rail against the king:

> "My lord," he shouted at the King in his loud and nagging way, "what is your trouble now? What more do you want? Your huts are full of bronze and . . . you have plenty of the choicest women in them too. Maybe you are short of gold. . . . Or one more girl to sleep with and to make your private property, though it ill becomes you as our general to lead the army into trouble through such practices." (*Iliad* ii, p. 46)

It was the institution of free debate more than anything else, I believe, that set Greece above all other nations.[4] A debate is a competition of minds, in which the contestants must counter one another with arguments designed to persuade their peers. The key words are *competition, argument, persuasion,* and *peers*—all aspects of what we mean by objectivity and, ultimately, science.

The techniques of debating were studied and taught from earliest times. As Phoenix reminds Achilles:

> Did not the old charioteer Peleus [Achilles' father] make me your guardian when he sent you off from Phithia to join Agamemnon [in the war against Troy]. You were a mere lad, with no experience of the hazards of war, nor of debate, where people make their mark. It was to teach you all these things, to make a speaker of you and a man of action, that he sent me with you. (*Iliad* ix, p. 172)

In Homer's time, the techniques were still rather crude, though colorful. Insults, as we have seen, were acceptable. But reference was generally made only to public knowledge, such as the number of women that Agamemnon had captured, rather than to private visions or conversations with a god. Appeals to omens and their interpretation were also used, because augury was an accepted practice at the time.

Augury, the practice of making predictions from omens, is a form of prophecy; but an augur isn't a prophet, at least not in the biblical sense that we have been using the term. A prophet is someone who claims to be an emissary of a god—that is, a holy man. An augur is someone who claims to be skilled in the interpretation of omens—that is, a clairvoyant. Omens are messages from the gods that anyone can see but that only someone with the gift of second sight can properly interpret.

Even more than war and debate, the *Iliad* and the *Odyssey* are dominated by the sea. Greece has always been a littoral culture with a maritime economy. Greek merchant ships traded as far west as the Pillars of Hercules (Gibraltar) and as far east as Scythia on the Black Sea. Greek settlements dotted the coasts of southern Italy, Sicily, North Africa, and Asia Minor, and Greek trading posts were established in many foreign ports. This made travel relatively easy for anyone seeking adventure or fleeing persecution. The *Odyssey* exalts the resourcefulness of the adventurer as much as the *Iliad* exalts the heroism of the warrior.

The sea is freedom, adventure, wealth, and knowledge—all factors important to the development of science. With the many Greek-speaking city-states on the coasts and islands of the Mediterranean, a dissident could easily find freedom in a rival city. Homer regales us with stories of young men who flee one city to start life over in another. Odysseus has for almost 3,000 years epitomized the solitary adventurer. Washed naked onto the shores of Phaeacia, he must survive on his wits alone.

A prosperous mercantile trade created a merchant class that could afford education for its children (Bowen, 1972). The importance of this can't be overstated because it created the teaching profession. Secular schools, free from religious or civil control, arose to meet the needs of the wealthy burghers. In classical times, the more famous teachers, such as Protagoras, went from city to city giving advanced seminars. At one such seminar, Protagoras explains the educational system to Socrates:

> Mother and nurse and father and tutor are vying with one an-
> other about the improvement of the child as soon as ever he is
> able to understand what is being said to him. . . . They send
> him to teachers, and enjoin them to see to his manners even
> more than to his reading and music; and the teachers do as they
> are desired. (Plato, *Protagoras;* in Buchanan, 1948, p. 67)

The technology of shipbuilding and the demands of a seagoing
voyage have their own logical imperatives that require linear, reality-
oriented thinking. There are no mystics on board ship. Homer gives
us this appreciation of the shipwright's craft:

> Odysseus began to cut the timber down. . . . Twenty trees in
> all he felled, and lopped their branches with his axe; then
> trimmed them in a workmanlike manner and trued them to the
> line. . . . With [augers] he drilled through all his planks, cut
> them to fit across each other, fixed this flooring together by
> means of dowels driven through the interlocking joints, giving
> the same width to his boat as a skilled shipwright would
> choose in designing the hull for a broad-bottomed trading ves-
> sel. He next put up the decking, which he fitted to ribs at short
> intervals, finishing off with long gunwales down the sides. He
> made a mast to go in the boat, with a yard fitted to it; and a
> steering-oar too, to keep her on her course. . . . [Sails] he man-
> ufactured too; and then lashed the braces, halyards, and sheets
> in their places on board. Finally he dragged her down on roll-
> ers into the tranquil sea. (*Odyssey* v, p. 94)

The cosmopolitanism of the seagoing Greeks contrasts sharply
with the parochialism of the desert-dwelling Hebrews. There is almost no mention of the sea in the Old Testament, and when there
is, it is as a metaphor for destruction: "Save me, God, for the waters
have closed in up to my neck. I am sinking in the deepest swamp
and there is no firm ground. I have stepped into deep water and the
waves are washing over me" (Ps. 69:1–2). The only seagoing voyage in the Old Testament is in the story of Jonah. This is a short
cautionary tale of late postexilic origin, in which the multinational
crew of the ship and Jonah's journey to Nineveh symbolize the expanding foreign contacts of the Jews of the Diaspora.

The assembly and the sea were central factors in Greek life long
before Homer, but the Olympian religion of the *Iliad* and the *Odyssey* was his invention. That is, whatever religion existed in Homer's
time, succeeding generations knew only Homer's. Just as the editors
of the Old Testament had the final say about Yahweh, Homer had

the final say about the Olympians. And because Homer was a popular poet and not a priest, his religion is joyous, without sin or piety.

If Yahweh is the projection of an absolute desert patriarch, Zeus is the projection of a loving Greek family man. His immediate family, the Olympians, consists of his wife, Here, his mistress Leto, and various sons and daughters by these and other goddesses. It's a noisy, quarrelsome family, whose meetings reflect the openness of the Greek assembly.

After Agamemnon takes Briseis from Achilles, Achilles calls on his mother, Thetis, a sea nymph, to ask Zeus to help the Trojans push the Achaeans back to the ships and slaughter them. She agrees but must wait twelve days before she can see Zeus, because all the Olympians are away at a banquet with the worthy Ethiopians. Heaven is closed for vacation! This comic touch is typical of how Homer contrasts the dark conflicts of the humans with the light-hearted affairs of the gods.

Thetis does succeed in persuading Zeus to promise to help the Trojans, which gets him in trouble with his wife, Here, who favors the Achaeans. A heated quarrel ensues, which Zeus ends by making godlike threats. The tension is broken when Here's crippled son Hephaestus runs around serving sweet nectar to all the gods, causing them to break out laughing. A merry banquet follows, after which Zeus, "the Lord of the Lightning Flash," retires with Here beside him.

The gods portray life at its fullest, whereas humans portray life at its most heroic. The gods often smile and laugh; they're beautiful and immortal. The humans are in deadly earnest, fighting and dying in the dust of the Trojan plain. The gods don't expect love, reverence, or obedience from humans, but they do enjoy a juicy burned offering. They interfere obsessively in human affairs, not to impose some moral order, but to amuse themselves.

Homer's religion has no priests or pretense. The gods aren't inscrutable; Homer reveals their every thought and emotion. Here, intent on circumventing Zeus's order that she and the other gods stop interfering in the war, tucks a love potion into her bosom and goes to seduce him away from his vigil on Mount Ida, where he is viewing the battle in peace. As she approached,

> Zeus the Cloud-compeller saw her, and at the first look his
> heart was captured by desire, as in the days when they had first

enjoyed each other's love, and gone to bed together without their parents' knowledge. . . .

"Here," said the Cloud-gatherer Zeus, ". . . let us enjoy the delights of love. Never has such desire, for goddess or woman, flooded and overwhelmed my heart. . . ."

"Dread Son of Cronos, you amaze," said the Lady Here, still dissembling, "Suppose we do as you wish and lie down in each other's arms on the heights of Ida where there is no privacy whatever, what will happen if one of the eternal gods sees us asleep together and runs off to tell the rest?" (*Iliad* xiv, pp. 64-65)

Zeus, of course, prevails, just as Here planned he would. And as fresh grass grows beneath them and a beautiful golden cloud covers them, the other Olympians are free to jump into the fierce battle raging below.[5]

To be sure, this charming scene, delightful in itself, is a literary device used to highlight the horrors of the ensuing battle. But it is also a part of Homer's poetical cosmography. Unlike the Old Testament, which looks up to God in awe and fear, Homer looks down on his gods from some great cosmic perch. It isn't that the gods aren't real; on the contrary, it's because they're so real, so finite, that Homer can treat them as characters in his story.

The concept of finite gods is alien to the Judeo-Christian mind, but it is eminently logical. It provides humans with ample outlets for their propensity to seek the help of the supernatural while explaining why the supernatural so often fails them. It creates no contraction between the gods and natural law, because the gods are subject to natural law.

Homer makes this explicit in the standard response that one god gives another when asked for a favor: "I shall gladly do what you ask of me, if I can and if it is not impossible"(*Iliad* xiv, p. 262). To the egocentric mind, nothing is impossible. As Jesus put it, "In truth I tell you, if your faith is the size of a mustard seed you will say to this mountain, "Move from here to there," and it will move; nothing will be impossible for you" (Matt. 17:20). Belief in impossibilities is the starting point for logic, deductive mathematics, and natural science. It can originate only in a mind that has freed itself from belief in its own omnipotence.

Homer, of course, was a poet, not a logician. When he has a god give the "if it is not impossible" caveat, he may be criticizing rival poets who composed stories in which gods could travel instantaneously, revive the dead, or go back in time. Homer's gods do none of these. In fact, the most fanciful parts of the *Odyssey*—Odysseus' adventures with the Cyclops and with Circe, his trip into Hades, and his encounters with Scylla and Charybdis—are told by Odysseus himself to his credulous Phaeacian benefactors who are all too eager to believe they are hosting a true mythic hero. Homer seems to have deliberately distanced himself from these tall tales, putting them into the mouth of nimble-witted Odysseus, whose ability to deceive with a false story is demonstrated elsewhere in the epic.

Greek culture must have had a penchant for realism and noncontradiction even before Homer. But just as Galileo, through the power of his intellect and the skill of writing, pushed current ideas about mechanics to new heights, so Homer advanced the predilections of his time and enshrined them in his masterpieces for all ages.

Greek Rationality

The development of objective thinking by the Greeks appears to have required a number of specific cultural factors. First was the assembly, where men first learned to persuade one another by means of rational debate. Second was a maritime economy that prevented isolation and parochialism. Third was the existence of a widespread Greek-speaking world around which travelers and scholars could wander. Fourth was the existence of an independent merchant class that could hire its own teachers. Fifth was the *Iliad* and the *Odyssey*, literary masterpieces that are themselves the epitome of liberal rational thinking. Sixth was a literary religion not dominated by priests. And seventh was the persistence of these factors for 1,000 years.

That all these factors came together in one great civilization is quite fortuitous; it didn't happen twice. Yet Greece's unique discovery of objective thinking didn't inevitably lead to abstract deductive thinking. Had the Persians defeated Alexander at the Battle of Issus (333 B.C.), Europe might have been Orientalized instead of Asia's becoming Hellenized. Then some prophetic religion out of Palestine could have spread across Europe 400 years before Christianity. We might have had Socrates, but not Euclid.

Athens and Jerusalem, only 800 miles apart, have always represented very different modes of thinking. Greek culture, with its penchant for reason and reality, very early broke through the barrier of egocentrism. Then, as though a giant blindfold had been removed from the eyes of humanity, the external world was seen for the first time. It isn't a sight that pleases everyone.

In the second century, the Christian writer Tertullian dismissed objectivity altogether with the exclamation, "What has Athens to do with Jerusalem?" And even though Thomas Aquinas tried to mate Aristotelian logic with church dogma, the inherent incompatibility of reason and faith made the union sterile. This conflict is still with us today, as traditional Muslims try to reconcile their ancient ways of thinking with the demands of modernity and secularized American Catholics try to reconcile their ideal of free conscience with the dictates of a dogmatic church.

But Alexander did win the Battle of Issus. Eighteen months later, he founded the city of Alexandria, which for the next 700 years was a major center of Greek learning. There mathematics and astronomy advanced to a height not surpassed for another 1,300 years. Central to these achievements was the discovery of deductive reasoning sometime before 600 B.C. It is to this epochal event and its stunning consequences that we go in Chapter 5.

5

Theorems and Planets

Early Mathematics

Some years ago, a student asked me why he had to prove things in geometry. "After all," he said, "I learned in sixth grade that the sum of the angles of a triangle equals 180 degrees. I believe it. Why do I have to prove it?"

I was shocked. For over 2,000 years, geometry has been taught to students in order to develop their ability to think abstractly. For some, understanding the first Euclidean theorem is the entrance into a new world of higher thought in which the mind, freed from the bounds of illusive words, first comprehends pure logic. For others, geometry is just a dreary and confusing mumbo jumbo of meaningless theorems to be painfully memorized.

I tried to answer the student's question as best I could, discoursing on the difference between empirical and deductive knowledge and the importance of mathematical rigor. My answer couldn't have been satisfactory, however, since he soon left science for religion and economics.

The abstract reasoning involved in a mathematical proof goes well beyond the reasoning of ordinary life. The human short-term, or working, memory is very small, between four and seven chunks (see Chapter 3). The most complex reasoning—whether by a chess grand master or a theoretical physicist—is done while thinking about only a few items, or chunks, at a time. The master differs from the novice by how much information each chunk can carry and how the chunks are organized in long-term memory.

An Early Theorem

To make this point clear and to demonstrate that abstract thinking is a major advance over ordinary thinking, I will give a proof of one of the oldest theorems in mathematics. From an inscription found on a clay tablet, we know that the Babylonians were familiar with it in 1700 B.C. The proof requires some background knowledge, which is given in the next three paragraphs. This will be familiar to some readers and unfamiliar to others. It isn't necessary that you understand the next three paragraphs, but it is necessary that you appreciate how much information is behind words such as *square* and *square root*.

All lengths can be expressed as a multiple of some basic unit length. To be specific, let the unit of length be 1 inch. Then in terms of this unit, my height is 68 inches; the circumference of the earth is 1.58 billion inches; and the diameter of an atom is 0.000,000,002 inch. In a similar way, all areas can be expressed as a multiple of a basic unit area. We can define this unit to be equal to the area of a square that is 1 inch on a side. We call this unit the *square inch*. It is clear from Figure 5.1 that a square 2 inches on a side contains exactly four unit squares and so has an area of 4 square inches. Like-

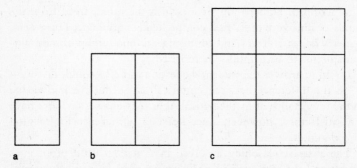

Figure 5.1 (a) One square inch is a unit of area equal to that of a square that is one inch on a side. (b) A square that is two inches on a side has an area of four square inches. (c) A square that is three inches on a side has an area of nine square inches.

wise, a square 3 inches on a side contains exactly nine unit squares and so has an area of 9 square inches.

These are examples of the general rule that the area of a square is obtained by multiplying the length of its side by itself. The area of the 2-inch square is $2 \times 2 = 4$ square inches, and the area of the 3-inch square is $3 \times 3 = 9$ square inches. For this reason, a number multiplied by itself is called the *square* of the number; 4 is the square of 2, or 2 squared, and 9 is the square of 3, or 3 squared. The square of any number a is $a \times a$, and is written a^2 (read "a squared"). For example, $2^2 = 2 \times 2 = 4$, $3^2 = 3 \times 3 = 9$. In this notation, the area A of a square with sides a is written $A = a \times a$, or $A = a^2$.

The reverse problem is to find side a of a square of a given area A. Clearly, a square with an area of 16 square inches has sides of 4 inches, since $4^2 = 16$. We call 4 the *square root* of 16, and write it $\sqrt{16}$. By definition, the square root $\sqrt{A}$ of any number A is the number that, when squared, equals A. Thus $\sqrt{16}$ is 4 because $4 \times 4 = 4^2 = 16$. Likewise, the square root of area A of a square is equal to the side of the square. That is, $a = \sqrt{A}$, since $\sqrt{A} \times \sqrt{A} = A$.

The last three paragraphs show just a little of what is behind the symbols a^2 and $\sqrt{A}$. It probably takes months, if not years, for a person to become so familiar with these symbols that they can be used in working memory as single chunks to represent all the information of the preceding paragraphs. If you have such familiarity, then it will be easy to follow the proof I am about to give. But if you don't, the proof will be almost incomprehensible. You will constantly have to go back to the preceding discussion to check the meaning of terms, and in the process you will lose the main line of argument. The process of understanding even this simple proof exceeds the capacity of the working memory of anyone not already familiar with most of the terms.

Theorem: The diagonal of a square is $\sqrt{2}$ times the length of its side.

Figure 5.2a shows a square with sides a and diagonal b. I claim that $b = \sqrt{2}a$.

Proof: The area of this square is $A = a^2$. A second square, with sides b, is drawn on the diagonal (Figure 5.2b). The area of this second square is $B = b^2$. Area A is composed of two equal triangles, and area B is composed of four of the same triangles. Therefore, area B is twice area A, or $b^2 = 2a^2$. The square root

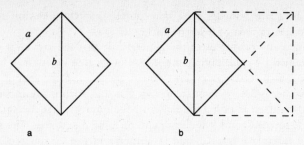

Figure 5.2 (a) A square with sides *a* has a diagonal *b* and an area a^2 equal to two triangles. (b) A square with sides *b* has an area b^2 equal to four triangles. Thus $b = \sqrt{2}a$.

of b^2 is *b*, by definition, and the square root of $2a^2$ is $\sqrt{2}a$. This proves that $b = \sqrt{2}a$.

The level of reasoning required for this simple proof is considerably higher than that required for ordinary thinking. There is nothing natural about it. Such a higher level is obtained through formal education extending over many years. Not only must the student become completely familiar with the precise meaning of many technical terms—*square, square root, diagonal*—and of a number of implicit axioms— the square root of equals are equal—but he must learn how to represent each idea by a single chunk so that several ideas can be considered at the same time.

Although not particularly rigorous, this proof exemplifies the fact that every proof is an argument designed to convince a critical respondent. It does this by starting with statements that both the author and the reader accept as true; for example, a square with sides of length *a* has the area a^2. It then proceeds to build a logical case for its conclusion from the commonly accepted premises.

Early Greek Mathematics

This procedure for formulating a convincing argument seems so natural to us that it is difficult to believe that it is a unique invention of the ancient Greeks. Although the origin of Greek mathematics is lost in time, the earliest name associated with Greek mathematics is that of Thales of Miletus (624–547 B.C.), one of the Seven Wise Men. Plutarch tells us "he was apparently the only one of these [Wise Men] whose wisdom stepped, in speculation, beyond the limits of practical utility," and Pliny tells how Thales "measured the heights of pyramids and all other similar objects by measuring the shadow of the object when a body and its shadow are equal in length" (Heath, 1921, vol. 1, pp. 128–29)—that is, when the sun is such that a stick set upright in the ground casts a shadow equal to its height, the shadow of the pyramid will be equal to the pyramid's height.

Shortly before 500 B.C. the Greek mathematician Pythagoras (572–497 B.C.) discovered that the square root of 2 is an irrational number. This event is both a metaphor for and an example of the abstract deductive mode of human thought that the Greek mathematicians were inventing. As we shall see, it is the kind of discovery that is possible only once one has obtained a level of abstraction beyond that of ordinary thought.

Little is known for certain about Pythagoras, since no writings of his have survived. He is said to have traveled and studied widely, spending many years in Egypt, perhaps after studying with Thales himself for a time. Later, in the Greek colony of Croton on the southern coast of Italy, he founded a mathematical–mystical cult dedicated to mathematics and asceticism. No one remembers much of the philosophy of the Pythagoreans (something about "everything is numbers" and never eating beans), but their mathematical accomplishments are the foundation of all subsequent mathematics and science (Sarton, 1960).

What Pythagoras discovered 2,500 years ago is that it is not possible to measure exactly the diagonal of a square with any ruler that exactly measures a side. To understand the problem better, consider a square with sides 1 inch long. According to the theorem proved earlier, the diagonal is $\sqrt{2}$ inches. Figure 5.3a shows a measurement of this diagonal with a ruler graduated in quarter inches. Such a

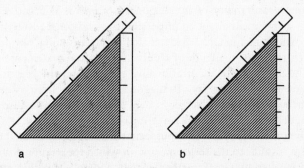

a   b

Figure 5.3 Measurements of the diagonal and side of a square with one-inch sides, (a) with a ruler graduated in quarter inches and (b) with a ruler graduated in eighth inches.

ruler measures the 1-inch side exactly, since four quarter-inch divisions fit exactly into 1 inch. But this ruler doesn't measure the hypotenuse exactly, since the upper edge of the diagonal lies between the 5-quarter and 6-quarter mark on the ruler. The length of the diagonal is between 1.25 inches and 1.50 inches, and so $\sqrt{2}$ is between 1.25 and 1.50.

We can try using a ruler with smaller divisions, say eighth inches. But as Figure 5.3b shows, the upper edge of the diagonal still lies between two divisions, here between the 11-eighth and 12-eighth mark, and so $\sqrt{2}$ is between 1.375 and 1.50.

Of course, there is nothing unusual about not being able to measure a length exactly with a ruler graduated in quarters or eighths. What Pythagoras discovered was that no matter what divisions are used to divide the inch—whether sixteenths, thirty-seconds, hundredths, thousandths, millionths, or whatever—one end of the diagonal will always fall between two divisions. This means that $\sqrt{2}$ must be represented by an infinite number of terms that never repeat; we write $\sqrt{2} = 1.414213\ldots$, the ellipsis dots indicating that the decimal goes on forever. The idea that there were lengths that couldn't be measured exactly, even in principle, was so baffling that they were labeled irrational.

The irrationality of $\sqrt{2}$ is one of the earliest discoveries in mathematics that could not have been anticipated by concrete examples. If you draw a right triangle and measure the lengths of its sides, you will find that the square of the hypotenuse equals the sum of the squares of the other two sides, at least approximately. Thus the Pythagorean theorem was certainly known before it was proved. On the other hand, an approximate measurement of the diagonal of a square gives no hint that the diagonal can't be measured exactly. The nonmeasurability of the diagonal is a universal negative statement—a statement about an impossibility—and consequently has no approximate form. Only abstract deductive reasoning can prove—or even suggest—the truth of a universal negative statement.

The discovery of the irrationality of $\sqrt{2}$ upset the very foundations of Greek geometry. Pythagoras is said to have sworn the members of his sect to secrecy. One story has it that a disciple of Pythagoras, the mathematician Hippasus, was drowned at sea for revealing this discovery. Although no truth can be attributed to this story, it testifies to the seriousness with which the irrationality problem was taken by Greek mathematicians.[1] Plato wrote that men unacquainted with the existence of rational numbers are more comparable to swine than to human beings,[2] meaning, perhaps, that for Plato too the understanding of irrational numbers represented a critical level of human reasoning.

The irrationality of $\sqrt{2}$ is interesting, not only because of its great historical importance, but also because it's such a peculiar idea when you first think about it. The proof of the irrationality of $\sqrt{2}$ begins by defining a rational number as any number that can be written as a fraction of two whole numbers. Clearly 1/2, 2/3, and 3/4 are rational. So are 105/16 and 524879/16943. Even 1, 25, and 106 are rational, since they are equal to 1/1, 25/1, and 106/1. If $\sqrt{2}$ were a rational number, say 141/100, then a ruler divided into hundreds of an inch would measure the diagonal of a 1-inch square to be exactly 1.41 inches.

The proof that $\sqrt{2}$ is irrational is by contradiction. That is, one assumes that there are two whole numbers n and m, such that $\sqrt{2} = n/m$, and then shows that this leads to a logical contradiction. The proof[3] is worth studying not only for its historical interest, but also because it's a beautiful example of the clarity and power of abstract deductive mathematics. With this proof by Pythagoras, humankind for the first time in history made a discovery by pure reason that could not have been anticipated by intuition.

Egyptian and Babylonian Mathematics

Of course, Greece didn't "invent" mathematics. The older civilizations of Egypt and Babylonia had mathematics for thousands of years before Greece did. But knowledge of this earlier mathematics was lost for more than 2,000 years. Only since the nineteenth century, when the writings of ancient Egypt and Babylonia were deciphered, have we begun to understand the mathematics of these early civilizations.

Egyptian mathematics, it turns out, is very elementary and was probably surrounded by an aura of magic. The Rhind Papyrus, an Egyptian arithmetical scroll from the seventeenth century B.C. calls itself "accurate reckoning. The entrance into knowledge of all existing things and all obscure secrets." It is a copy of a much older work, strengthening the view that the scribe believed he was dealing with ancient occult knowledge. In fact, the scroll is just a compilation of arithmetic problems (Chase, Manning, and Archibald, 1927).

Babylonian mathematics, on the other hand, turns out to be unexpectedly sophisticated. The Babylonians wrote on clay tablets, the only writing medium that fire preserves rather than destroys. Consequently, in the basements of museums around the world, dust collects on hundreds of thousands of Babylonian tablets unearthed from ancient archives destroyed by fire. Of course, most of these tablets concern matters other than mathematics, such as business, law, and religion (Pritchard, 1958). Only a few hundred concern mathematics, and these are mostly from the time of Hammurabi (ca. 1700 B.C.).

One of these, YBC+7289, in the Yale Babylonian Collection, shows a square with one diagonal drawn (Aaboe, 1964). Numbers indicating the lengths of a side and the diagonal clearly show that the Babylonians knew that the diagonal was $\sqrt{2}$ times a side. They even knew that $\sqrt{2}$ is approximately 1.4142. There is no indication, however, that the Babylonians ever realized that $\sqrt{2}$ is irrational.

Most of Babylonian mathematics involved algebra problems—that is, problems in which one must find an unknown number that satisfies some given condition. Here's an example from a tablet in the British Museum (BM13901):

> *Problem:* A square and two-thirds of its side is 7/12. What is the side of the square?

> *Solution:* Half of 2/3 is 1/3. Square this and add to 7/12 to yield 25/36. Take the square root of this and subtract 1/3 from it. The result is the solution, 1/2. (After Aaboe, 1964, p. 23)

This problem is equivalent to finding a number x that satisfies the equation $x^2 + 2/3\ x = 7/12$. Equations such as this, which contain the square of the unknown, are termed quadratic. In high-school algebra, students are taught the general solution to quadratic equations, but Babylonian mathematics was concerned exclusively with solving specific problems. No proof or explanation of a general method was ever given. However, the great variety of problems found, some quite involved, indicate that a general method was being taught, if only by example.

Babylonian mathematics appears to be the most sophisticated ever reached independently of the Greek tradition. As Otto Neugebauer, the great interpreter of Babylonian tablets, wrote:

> [With Babylonian mathematics] we are dealing with a level of mathematical development which can in many respects be compared with the mathematics, say, of the early Renaissance. Yet one must not overestimate these achievements. In spite of the numerical and algebraic skill and in spite of the abstract interest

> which is conspicuous in so many examples, the contents of
> Babylonian mathematics remained profoundly elementary. In
> the utterly primitive framework of Egyptian mathematics the
> discovery of the irrationality of $\sqrt{2}$ would be a strange miracle.
> But all the foundations were laid which could have given this
> result to a Babylonian mathematician, exactly in the same arith-
> metical form in which in which it was obviously discovered so
> much later by the Greeks. (1957, p. 48)

Babylonian mathematics was at its height in 1700 B.C. and continued for another 3,000 years without further development. That
is, the mathematics on tablets from 300 B.C. is indistinguishable from
that on tablets from 1700 B.C. and the Arabs continued this oriental
style mathematics almost into the modern era.

This lack of development makes it questionable whether, after
the innovative phase, the term *mathematics* should be applied to this
sort of problem solving. It is more akin to solving brain teasers;
although obviously above the ordinary level of thinking, it doesn't
involve the level of abstract deductive reasoning required to prove
that $\sqrt{2}$ is irrational.

Deductive Mathematics

Only in the eleventh and twelfth centuries did algebra begin to use
the deductive methods of Greek geometry. This in time led to modern algebra, which is an enormously efficient method for drawing
the logical implications of complex relations. To this day, our educational system retains the memory of the independent developments of algebra and geometry by teaching them as separate subjects.

Thus although the antecedents of Pythagoras can be traced back
through the Ionian philosophers to still earlier Egyptian and Babylonian sources, Pythagoras and the Greek mathematicians of the following few centuries created a level of abstract deductive mathematics never reached by any other peoples. The Egyptians, for
example, knew that a triangle with sides 3, 4, and 5 was a right
triangle, but they did not know the general Pythagorean theorem.
But what does it mean to know a geometrical "fact"? A carefully
drawn triangle with sides 3 and 4 will have a hypotenuse that measures 5 to within the accuracy of the drawing. But no drawing of a
triangle, no matter how accurate, can be exact. The geometer's triangle is composed of three straight lines that are without breadth.

Real lines drawn on paper or papyrus have breadth. The geometer's lines meet at a point that has neither length nor breadth.

Geometric points and lines are abstractions that have only those properties that geometers, through their definitions and postulates, say they have. Geometers, Plato observed, "make use of visible figures and argue about them, but in so doing they are not thinking of these figures but of the things which they represent" (*Republic,* bk. vi; quoted in Heath, 1921, vol. 1, p. 290). It is only by analogy that the postulated properties of geometric points and lines correspond, more or less, to those of real points and lines (at least in Euclidean geometry). Indeed, most of us would have an impossible task trying to understand a geometric argument without a diagram. But this is a limitation of our working memory, not an acknowledgment that the diagram is the thing it represents. With modern algebraic notation, mathematicians can prove theorems concerning four- and higher-dimensional spaces for which diagrams aren't possible.

This point is clear from our discussion of the measurement of the hypotenuse in Figure 5.3. Although the diagrams are a concrete representation of what we are talking about, the gap between these pictures and the abstract concept of an infinitely divided ruler must be crossed with a mighty leap of the imagination. It is the leap from concrete to abstract thinking, the leap from non-Greek to Greek mathematics.

Greek mathematics flourished for over 800 years, from the first primitive theorems concerning circles in the time of Thales (fl. 600 B.C.), through the completion of advanced work on ellipses and hyperbolas (conic sections) by Apollonius of Perga (fl. 200 B.C.), to the great astronomical work of Ptolemy (fl. A.D. 150).

The Pythagoreans are credited with putting geometry on the sound basis of theorems proved by logical deductions from definitions, postulates, and axioms, but it was Euclid who, some 200 years later, organized the results of Greek geometry into a comprehensive whole. His immortal work, the *Elements,* had no rival. All subsequent abstract mathematics has been done by mathematicians who learned, directly or indirectly, from it. The religions of the world have their various holy books, the Vedas, the Torah, the Gospel, the Koran, but for mathematics there is only the *Elements.*

The Hellenistic age, which followed the conquests of Alexander the Great (334–323 B.C.), spread Greek culture throughout the civilizations of western Asia. The vast Hellenistic empires that succeeded Alexander were run by Greek expatriates who brought with them their sporting contests and their marketplaces, their theaters and their schools. Greek scholars and teachers traveled as far east as the Indian border, and from there we may assume that Greek ideas, if not Greek traders, traveled along the trade routes into India and China.

The city of Alexandria, under the patronage of the Ptolemaic dynasty, which succeeded Alexander in Egypt, became a major center of Greek scholarship. Mathematicians, astronomers, geographers, and translators from many lands worked there to compile the great library of Alexandria. Even after the former territories of Alexander came under Roman rule in 30 B.C., Alexandria continued to nurture Greek learning. It was there that Euclid (fl. 300 B.C.) wrote the *Elements* and Ptolemy wrote his astronomical masterpiece, the *Almagest*.

Early Astronomy

The thousands of stars visible in the night sky form a fixed pattern. Different parts of the pattern are visible at different seasons of the year, and on any one night the pattern as a whole rotates about a fixed point that is close to Polaris, the North Star. But with only five exceptions, the stars don't move with respect to one another. These exceptions—Venus, Mercury, Mars, Saturn, and Jupiter—are the planets visible to the naked eye. Although seldom seen in the pale skies of our illuminated cities, they were startling apparitions in the pitch-black heaven of antiquity. Changing their positions from night to night, these five wanderers execute complex trajectories over time.

The problem of planetary astronomy is to predict these trajectories. Mars, for example, normally drifts eastward by about half a degree a night, relative to the background stars. But every 780 days, it reverses direction and drifts westward for about two months before reversing again and continuing its eastward course. Figure 5.4 shows the motion of Mars from October 1992 through May 1993 as it moved against the stars of the constellation Gemini. Its motion was reversed, or retrograde, from late November 1992 until mid-February 1993. Mars is again retrograde 780 days later, or from mid-January through late March 1995, and is then moving against the stars of the constellation Cancer. Jupiter and Saturn behave similarly, with their periods of reverse motion recurring every 399 and 378 days, respectively.

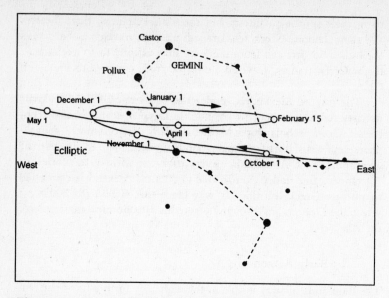

Figure 5.4 The path of Mars from October 1992 to May 1993. The motion was retrograde from late November to mid-February. (From William J. Kaufmann, III, *Universe*, 3rd ed. [New York: Freeman, 1991]. Reprinted by permission)

Many ancient peoples were keen observers of the planets' motion. The Babylonians and the Chinese kept extensive records of their observations that are useful to astronomers today, and tourists can still climb the beautiful Mayan–Toltec observatory (the Caracol) at Chichén Itzá. But it's important to realize that none of these peoples—for all the care and accuracy of their records—were doing astronomy. Rather, the purpose of the ancients' star watching was fortune-telling, pure and simple.

There are two basic forms of astrology. The more familiar horoscopic astrology correlates the positions of the sun, moon, and planets with events on earth. Therefore, the ability to predict the positions of astronomical bodies is the ability to predict human events. For this purpose, the Babylonians were experts in making accurate extrapolations from their extensive data. But they never tried to understand the motions underlying their observations.

The Chinese, on the other hand, for reasons discussed in Chapter 6, practiced portent astrology; that is, they looked for irregularities in nature—earthquakes, novae, eclipses—as signs from heaven.

Consequently, when they discovered that lunar eclipses followed a regular pattern, their interest in this phenomenon decreased (Nakayama, 1969).

Only the Greeks had the scientific interest and geometrical insight to realize that the earth is in nearly the same plane as the other planets, so that we observe the planets from an awkward perspective, looking outward from the center of the plane. Viewed from some Homeric perch far above this plane, the planets would be seen to be moving in more or less circular orbits.

Over the centuries, the Greeks developed many specific models of the solar system, both with the earth as the center and with the sun as the center. In the earth-centered, or geocentric, models, the sun and the planets orbit about a stationary earth. In the sun-centered, or heliocentric, models, the earth and the five planets orbit about a stationary sun. Aristarchus developed a tentative heliocentric model around 300 B.C. but it was not further developed. It was the geocentric model of Hipparchus (fl. 150 B.C.)—and, later, Ptolemy's improved version of it—that was universally used by astrologers and astronomers until the early seventeenth century.

A planetary model is a mathematical—in this case, geometric—description that is explicit enough to produce numbers that can be compared with observations. It differs from a theory in that no mechanism is proposed to explain why the planets should move as the model says they do. However, following Aristotle, it was universally believed that circular motion at constant speed (uniform circular motion), being "perfect," was the only motion suitable for celestial bodies. This is a mechanism of sorts, except that the observations can't be fit with a single circular motion. To explain retrograde motion, for example, it was necessary to move each planet in a small circle, the epicycle, about a center C that moved in a larger circle, the deferent, about a stationary earth (Figure 5.5). The motion around each circle is uniform.

In the case of Mars, point C, moving at a constant speed around the large circle, completes one revolution every 686 days. At the same time, Mars, moving at a constant speed around the small circle, completes one revolution every 365 days. These times were chosen to make the time between successive periods of retrograde motion equal to the observed value of 780 days.

The sun moves in its own circular orbit about the earth once every 365 days (actually 365.24 days), the same time that Mars and all the other planets move once around their small circles. These

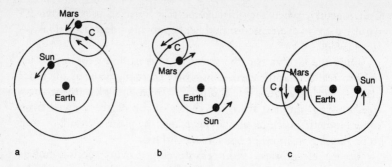

Figure 5.5 Geocentric system of Hipparchus for the earth and Mars: (a) Mars rotates about point C once every 365 days, and C rotates about the earth once every 686 days; (b) half a year later than (a), Mars is in the middle of its retrograde motion; (c) 780 days later than (b), Mars is again in the middle of its retrograde motion, but it is in a different position of its orbit.

coincidences of times were required to make the model fit the observations. Because a circle has 360 degrees, the sun moves just under 1 degree a day around its circle.

Likewise, Mars moves just under 1 degree a day around its small circle, and point C moves just over half a degree a day around the large circle. Thus, starting with known positions for the sun and Mars, their future positions, day by day, can be determined.

The mathematics required for these calculations isn't simple, for what astronomers actually observe isn't the position of Mars on these hypothetical circles, but Mars's angular position against the background stars as viewed from the earth. To calculate this angular position from the assumed positions on the circles requires trigonometry, a branch of mathematics invented by Hipparchus for this very purpose. This is discussed further in Chapter 6.

Hipparchus' planetary model, developed more than 2,000 years ago, is a particularly striking example of the uniqueness of the Greek mode of thought. Of all the civilizations that have closely observed the motions of the planets, only the Greeks thought to describe the complex irregularities of these motions by means of a geometric model. Moreover, they constantly compared the positions calculated by their models with their observations and modified their models accordingly. Hipparchus' model predicted the positions of the planets with an error of several degrees.

In the second century A.D., Ptolemy developed an improved

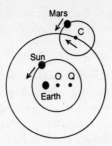

Figure 5.6 The geocentric system of Ptolemy for the earth and Mars. The earth is displaced from the center O of the large circle, and C moves with a constant angular speed relative to the third point Q.

version of this model that could predict planetary positions to better than 1 degree (Hoyle, 1975). In the Ptolemaic model, the earth was displaced slightly from the center of the large circle, and the angular motion of C, the center of the smaller circle, was uniform, not with respect to the earth or to the center of the large circle, but to a third point Q, on the opposite side of O from the earth (Figure 5.6). The scheme was complex but workable, and it dominated astronomical thought for the next 1,400 years.

Christianity

With the spread of Christianity in the third and fourth centuries, a liberal Greek education—which had marked an educated man (and occasionally a woman) for the previous 800 years—was increasingly seen as improper for Christians. In 415, Hypatia, an Alexandrian mathematician renowned for her learning and beauty, was brutally murdered by a band of monks working for the archbishop, St. Cyril. In 529, the Academy in Athens, founded by Plato nearly 1,000 years earlier, was closed by the ecclesiastical authorities for being too pagan.

In western Europe, the decline in learning was accelerated by the repeated attacks by the Goths, Visigoths, Vandals, and Huns. The Hun leader Attila invaded northern Italy in 452. Rome itself was taken in 476 by the Goth leader Odoacer, who deposed the last Roman emperor (Romulus Augustulus).

In the world that followed, there was little opportunity for studying geometry and astronomy, as these disciplines can be learned only through an expensive formal educational process involving time, effort, and trained teachers. Such educational institutions exist only when there is an elite wealthy and interested enough to invest the necessary time, money, and effort in study. In the new barbarian–Christian Europe, with its militaristic and ecclesiastical elites, abstract learning had a low priority. And so, with no market for teachers and scholars, schooling and scholarship didn't develop. Feudal Europe, after all, had no need for mathematics beyond that required in the marketplace. It isn't the absence of mathematics and science in a society that is surprising, but the fact that such learning has flourished at all in a few places from time to time.

The Dark Age that descended on Europe was intensified by the Arab conquest of North Africa and western Asia after 622, which cut the remaining trade links between Europe and the more developed societies to the east, thereby isolating the continent in its ignorance. The total dominance of Christianity, a mystical religion with an illogical theology, discouraged the development of logical thinking even (perhaps especially) among the few who could read. Compared with the intellectual brilliance of the Hellenistic age, the light really did go out in Europe's Dark Age.

We, living 500 years on the other side of the Renaissance, may see the gradual emergence of Europe from its ignorance as a sign of the inevitability of progress. But even though some opening to secular knowledge must sooner or later have occurred, the scientific revolution that ensued can be traced to the flame of Greek mathematics that were kept burning in the East. And the survival of this knowledge was anything but inevitable.

Constantinople remained the capital of the Byzantine Empire after the fall of Rome. Roman in political structure, Greek in language, and Christian in religion, it continued as a major center of Christian learning for over 1,000 years. However, religion increasingly dominated its intellectual life, and although the ancient works of Euclid and Ptolemy were preserved in its libraries, fewer and fewer students studied them.

Arab Science

The Hegira, the flight of Muhammad from hostile Mecca to Medina in 622, marks the beginning of Arab ascendency in the East. From Medina, where he established his theocratic state, he set out to win converts. So successful was Muhammad at this that in eight years his army was able to take Mecca without a fight. Muhammad died two years later, but under the banner of Islam his armies swept westward into North Africa and eastward into Persia and Central Asia. By 642, the Arabs had captured Alexandria.

Originally a band of uneducated tribesmen, the Arabs soon absorbed the high culture of the lands they captured. In 762, Caliph al-Mansur (the victorious) founded Baghdad, his magnificent new capital on the banks of the Tigris River (about sixty miles from the ruins of Babylon). It is said that 100,000 workers built the city in four years.

Around 770, a scholar from India brought to Baghdad a set of Hindu astronomical works, called the *Siddhantas*. These works used an early form of trigonometry and astronomy, which was probably transmitted to India during the Hellenistic period. Now India was repaying its debt with interest, because these works also introduced the Hindu decimal system to Islam, from which it eventually reached the West. Fascinated by what he must have thought was an advanced form of astrology, al-Mansur arranged to have the *Siddhantas* translated from Sanskrit into Arabic. When he learned that still more ancient works existed in Greek, he obtained copies of the *Elements* and the *Almagest*[4] from Constantinople. With their translation into Arabic, Arabic began to replace Greek as the language of learning in North Africa and western Asia.

Under the patronage of a succession of liberal caliphs, Baghdad became a great cosmopolitan center of learning, as Alexandria had been 1,000 years earlier. Scholars came from all parts of the empire and beyond—Greeks, Jews, Christians, Persians, Arabs, and Hindus—the same people who had contributed to learning in the Hellenistic period. Indeed, little changed except the language. The Arab rulers adopted the rich, opulent manners of their Persian subjects, and their scholars adopted the intellectual methods of the ancient Greeks. This was the golden age of Arab culture: the time of Scheherazade, the Arabian Nights, and the poems of Omar Khayyám. It lasted for more than 400 years and then ended as it had begun: In 1236 the Christians captured Cordova, the Islamic cultural center in the West, and in 1258 the Moguls captured Baghdad itself. In the aftermath, religion replaced science as the central intellectual force in Islam. But by then, Euclid and Ptolemy had been translated into Latin.

In the Arab Empire, the authority of the caliph was absolute, and all learning was by his leave. Al-Mansur kept an executioner sitting beside his throne on a leather carpet (to protect the tiles from blood). Al-Mamun, the sixth caliph of Baghdad, believed there was no difference between reason and religious law, built the House of Learning to rival the museum and library of Alexandria, and threatened to execute his Christian physician Hunayn for refusing to prepare a poison for an enemy (Hoyt, 1975).

Medicine and pharmacology, with special interest in the subspecialty of toxicology, were actively pursued throughout the empire. Hospitals were established; the works of Hippocrates and Galen were translated; and many original treatises on diseases and drugs were written. The Islamic prohibition on dissection inhibited surgery—probably for the best—but methods for removing cataracts were devised. Ophthalmology received special attention because of the high incidence of eye disease in the Middle East. There was a corresponding interest in optics. In an influential book entitled *Optics,* the Egyptian scientist Ibn al-Haitham (Alhazen, ca. 965–1039) investigated the refraction of light, reflection from concave and parabolic mirrors, and the magnifying effects of spherical lenses. He also corrected the curious Greek notion, shared by both Euclid and Ptolemy, that vision is caused by rays going outward from the eye to the object. Yet for all this, the Arabs didn't invent eyeglasses.

Astronomy was another great interest of the Arabs. The ability of the Ptolemaic model to calculate the planets' future positions opened the possibility of long-range astrological predictions. But the model's accuracy decreased with time, and so it was always necessary to update the predictions by basing them on the most current observations. Thus, as with all civilizations, the Arabs built many observatories, all for astrological purposes. Few of these observatories outlived their founding patrons, however, perhaps because their successors didn't trust the the astrologers who had failed to predict their own often-bloody succession. The Arabs made no significant advance on the Ptolemaic model itself.

The most important Arab contributions to science were in mathematics. Muhammad ibn Musa al-Khwarizmi (fl. 825) wrote important works that, in their Latin translation, introduced the Hindu decimal system to Europe. From the Latinized version of his name (Algorithmi) we get the word *algorithm,* and from his work *Hisab al-jabr wal-muqabala,* the word *algebra.* This work follows in the ancient Babylonian tradition of treating specific equations without any symbolism or theoretical basis. Without a notation for negative numbers, the equation $x^2 + 10x = 39$ is treated separately from $x^2 + 21 = 10x$ (for both of which, incidentally, a solution is $x = 3$).

The work also gives rules for calculating volumes and areas, again in the oriental rather than the Greek fashion. Indeed, this section has been traced to a Jewish text from A.D. 150. But we now know, from the decipherment of the Babylonian tablets, that this sort of mathematics had been practiced in western Asia, almost without change, for nearly 3,000 years (Berggren, 1986). Perhaps nothing better demonstrates that abstract thinking won't inevitably be discovered by intelligent mathematicians working at their profession. Most professionals, throughout the ages, have been content to think only what they have been taught.

Only with the rediscovery of deductive thinking does progress once again appear in mathematics. This is seen in the *Optics* of Alhazen and in the work of the Persian poet, mystic, and mathematician Omar Khayyám (ca. 1040–1124), who solved cubic equations in the Greek fashion, as the intersection of two conic sections.

The Greek Legacy

Greek mathematics and astronomy were introduced to Europe in the twelfth century, just as interest in them had all but disappeared in the East. The subsequent developments are described in Chapter 7. For now, it is sufficient to note that when the major breakthrough came in planetary astronomy with Copernicus's heliocentric model (1543), the printing press was there to guarantee its wide distribution.

There is an unbroken tradition that stretches from Pythagoras to Copernicus. But was Copernicus, and the scientific revolution that followed him, inevitable? I think not. In the 1,300 years between Ptolemy and Copernicus, the interest in astronomy was dominated by astrology. Although this certainly stimulated the development of better methods of observation, there was no interest in understanding how the planets moved and no improvement on the Ptolemaic model. Astronomy, like mathematics, had a few short periods of innovation followed by long periods of stagnation.

Greek geometry and speculative thought were unique inventions, never duplicated by other cultures, even those that had engaged in some kinds of mathematics for thousands of years. Furthermore, the Greek modes of thought—the cognitive structures developed through the study of deductive geometry—are absolutely essential to the work of Copernicus and his successors. The astronomical data compulsively accumulated over thousands of years by the ancients are of little value unless they are seen as resulting from an intelligible pattern of planetary motions. This notion that the cosmos is ultimately understandable is natural only to a mind that knows the power possible with abstract deductive reasoning. And finally, deductive mathematics is something that requires a disciplined formal education to learn. This means schools, teachers, and students. In the many political and religious upheavals over the millennia, the institutions of higher learning have often suffered total annihilation. The barbarian looter and the religious zealot are alike in their penchant for destroying what they don't understand. The glory of Greek mathematics was almost lost in the turmoil of the ages, and if it had been totally lost, it probably would never have been reinvented.

What about the Islamic world? Having kept Greek learning alive until its reintroduction into Europe, why didn't science develop further there? Again, the answer shows the precariousness of knowledge and learning. It seems the imams (prayer leaders), for religious reasons, forbade printing in Muslim countries. Except for one publishing house that existed briefly in mid-eighteenth century Turkey, it wasn't until the nineteenth century that printing came to Muslim countries (Boorstin, 1985). Thus for nearly 400 years, Islam was cut off from the scientific revolution that was taking place in Europe, the revolution it had helped initiate.

The importance of the Greek contribution to our civilization is acknowledged by all scholars. Besides deductive geometry, the Greeks originated theater, sports, history, philosophy, democracy—indeed, it has been claimed, consciousness itself (Jaynes, 1976). Their accomplishments are unique because they didn't invent just new subjects for thought, but fundamentally new ways of thinking. Because our entire scientific and technological civilization can, without question, be traced back to this Greek origin, it is a speculative question of some significance whether, without the Greeks, abstract deductive thinking would have developed sooner or later anyway.

As stated, such a question is unanswerable and, in a sense, meaningless. All that can reasonably be asked is whether there is any evidence that other civilizations ever independently developed abstract deduction. At first blush, the answer is obvious. There is no rival *Elements*. No other civilization, unfamiliar with the Greek legacy, ever developed deductive mathematics. There are no Chinese or Indian textbooks of theorems and proofs that predate contact with the Greek legacy.

This isn't to say that China and India haven't had immense influence on the development of science. Many seminal technological innovations, from the magnetic compass to paper, can be traced to China, and India kept the candle of Greek learning burning in the centuries between the fall of Rome and the rise of Islam. But the idea of science itself—of objectivity—is a uniquely Greek contribution.

Chapter 6 describes the intellectual climates of traditional China and India and shows how each culture was, in its own way, incapable of developing the concept of objectivity.

6
Sages and Scholars

There are two key points about the development of science for which there is universal agreement. First, the work associated with Copernicus, Galileo, and Newton in the sixteenth and seventeenth centuries created the new intellectual order that we call science. Second, this European development was directly stimulated by the work of ancient Greek mathematicians, astronomers, and natural philosophers.

The natural inference to draw from this—that science is a unique historical phenomenon outside the usual course of human development—isn't, however, universally accepted. On the contrary, it is commonly thought that India and China also had scientific—or protoscientific—traditions that, though perhaps not up to Greek standards, would have in time led to true science had they not been overtaken by the West. This viewpoint is due to a misunderstanding of the research in this area and of the nature of science itself. It is also due, perhaps, to a thoughtless relativism that pervades much of academia. What, after all, is so special about Western civilization?

In fact, for much of its history the West was a backwater of civilization. The anti-intellectualism of Christianity destroyed virtually all traces of Greek rationality in Europe for 1,000 years. During this period, the torch of Greek thought was kept alive by scholars in India and Arabia. Furthermore, many of the inventions on which European power was ultimately based—from the wheelbarrow to gunpowder—were imported from China. For 1,000 years after the fall of Rome, Europe was economically, culturally, and technologically inferior to the civilizations of the East.

The West's debt to these civilizations is incalculable. There can be no serious study of Western civilization that ignores the fact that virtually every idea and invention that contributed to the scientific revolution was imported from the East.

But having said this, we can still ask whether China or India itself was ever on a track—albeit a slow one—to science. Did either ever develop a theory of knowledge that was conducive to science? The obvious answer is no, since in 3,000 years of dedicated learning, they didn't develop science. They did develop many elaborate speculative systems, not one of which is objectively verifiable.

Much has been written on the question of why China, with its great learning and inventiveness, didn't develop science. This question will be discussed in detail later. But as Einstein wrote in 1953:

> The development of Western science has been based on two
> great achievements, the invention of the formal logical system
> (in Euclidean geometry) by the Greek philosophers, and the
> discovery of the possibility of finding out causal relationships
> by systematic experiment (at the Renaissance). In my opinion
> one need not be astonished that the Chinese sages did not make
> these steps. The astonishing thing is that these discoveries were
> made at all. (Quoted in Needham, 1969, p. 43)

To appreciate just how astonishing science is, we shall examine how two of the most successful civilizations on earth, those of India and China, managed without it.

Mystical India

India has had a continuous tradition of religious speculative thought from at least 2000 B.C. to the present. The great Hindu epics—the *Ramayana* and the *Mahabharata*—honor a warrior for his knowledge of the Vedas, just as Homer honors a warrior for his skill in debate. Respect for learning and knowledge appears to be as ancient in India as in any other extant civilization.

The knowledge of the Vedas and Upanishads—the sacred writings of the ancient Hindus—is mystical knowledge. The oldest work of Hindu literature, the *Rig Veda,* is a collection of hymns of a ritualistic cult controlled by a priestly class. These rituals and sacrifices, perhaps originally intended only to persuade the gods, developed in later Vedas into magical formulas that compelled the gods.

All magic, as we stated earlier, is based on the egocentric identification of the self with the supernatural. Hindu speculative thought eventually developed the ultimate egocentric identification, that of one's soul with the One of the universe. When one understood this identification, one obtained perfect knowledge and absolute power. As Franklin Edgerton explained:

> The Upanisads seek to know the real truth about the universe, not for its own sake; not for the mere joy of knowledge; not as an abstract speculation; but simply because they conceive such knowledge as a short-cut to the control of every cosmic power. The possessor of such knowledge will be in a position to satisfy his any desire. He will be free from old age and death, from danger and sorrow, from all the ills that flesh is heir to. By knowledge of the One that is All, and by mystically identifying his own self with that One which is All, he has, like that One, the All in his control. Knowledge, true esoteric knowledge, is the magic key to omnipotence, absolute power. (*Bhagavad Gita*, p. 121)

Traditional Hindu thought, with all its intricacies, never broke free from egocentrism. Indeed, it seems to have become ever more enmeshed in it with time. It concentrated more and more on the soul, totally disregarding material nature and even human consciousness itself. In Hindu belief, the soul is immortal and, after bodily death, transmigrates to a new body with a new consciousness. The *Bhagavad Gita* put it this way:

> As leaving aside worn-out garments, a man takes other, new ones, so leaving aside worn-out bodies, the soul goes to other new ones.
> Swords cut not [the soul], fire burns him not, water wets him not, wind dries him not.
> Not to be cut is he, not to be burnt is he, not to be wet nor yet dried; eternal, omnipresent, fixed, immovable, everlasting is he. (*Bhagavad Gita*, ii, 22–24, p. 121)

At its highest level, Hindu philosophy disregarded the material world and, in extreme cases, denied its existence altogether. The best minds were encouraged to renounce the world and become hermits in the forest. Some became sages—*rishis*—who founded spiritual communities—*ashrams*—where the countless variations and elaborations of Hindu thought were produced.

All these doctrines rested on commonly held cultural and metaphysical concepts that were universally accepted without question. One of these was the caste system, which divided humankind into four major groups—brahmans, warriors, farmers and merchants, and serfs—with hereditary duties and obligations. The brahmans were the smallest and most prestigious caste, consisting of priests, hermits, teachers, and scholars. So even though India always accorded to scholarship and learning the highest regard, they were monopolies of a closed community. The strict caste laws prevented nonbrahmans from marrying a brahman or from even acting as a brahman. In the *Ramayana,* the hero Rama actually seeks out and kills a sudra (serf) for practicing a yogic meditation (Coomaraswamy and Nivedita, 1913/1967).

Although scholars may debate whether the *Iliad* is the work of a single author, there is no question that the *Ramayana* and the *Mahabharata* are composite works. For centuries, it was the practice for Hindu poets and philosophers to append their works anonymously to these epics, until each became a vast storehouse of Hindu legends, poetry, and philosophy. The *Bhagavad Gita,* India's best-known mystical poem, is inserted just before the climactic battle of the *Mahabharata.* The *Mahabharata,* 90,000 couplets long, is seven times the combined length of the *Iliad* and the *Odyssey.* It is, according to Romesh Dutt, as though "the religious works of Hooker and Jeremy Taylor, the philosophy of Hobbes and Locke, the commentaries of Blackstone and the ballads of Percy, together with the tractarian writings of Newman, Keble, and Pusey, were all thrown into blank verse and incorporated with the *Paradise Lost*" (*Ramayana & Mahabharata,* p. 324).

In addition to adherence to the caste system, all Hindu writers accepted a pessimistic view of life. Existence is wearisome, and since death is always followed by rebirth, it is perpetual. The aim of religion is to break this cycle of birth and death by becoming one with the Absolute. Sects differed in the details of how to achieve this goal—whether by meditation, austerities, devotion, or indifference—but all accepted that to reach oblivion was an arduous task requiring a lifetime of work. Their views of the material world ran the gamut from indifference to abhorrence.

Hindu thought can be characterized as being extremely egocentric and nonegoistic at the same time. This isn't a contradiction, for the words, as used here, have very different meanings. A religious Hindu, like a religious Christian, tries to rid himself of all desires and passions. Going further, an ascetic withdraws entirely from worldly matters. Through meditation, he tries to do nothing, to think of nothing, to be nothing—to become, in fact, egoless. At the same time, the Hindu believes that by doing so his *atman*—his soul, or essential self—will become one with the Absolute and will obtain perfect mystical knowledge of the universe. The belief that one can, by concentrating on the space between one's eyes, obtain ultimate cosmic power is egocentrism at its most grandiose.

In Hindu thought, knowledge means mystical knowledge, usually the specific doctrines of a cult, and generally to the effect that the material world is unimportant or nonexistent. Practical knowledge of the merchant or artisan is never allowed to interfere with Indian philosophizing, since the caste system prevents a brahman from practicing a craft as much as it prevents a craftsman from becoming a philosopher.

That neither science nor protoscience developed in such an intellectual and social environment isn't surprising. Indian civilization strongly demonstrates that human thought can develop and thrive for thousands of years without ever stumbling on the notion of scientific objectivity. Science isn't the inevitable outcome of serious philosophizing.

The issue is complicated by a variety of claims that are made for the importance of Indian contributions to science. At one extreme, it is said that all science—indeed, all human thought—is contained somewhere in the Hindu scriptures (Coates, 1972). These are subjective interpretations by the devoutly religious, which the skeptic has no reason to accept. At the other extreme, it is said that Indian books of the eighth century helped initiate the great flowering of Arab science in the eighth century. These are historical interpretations by competent historians, which there is no reason to doubt.

But the fact that India was an important link in the chain of transmission of Greek science to the West doesn't answer the question of whether India ever initiated any steps toward science on its own. The question is further complicated by the conquests of Alexander, who in 327 B.C. extended Greek influence to the Indian states west of the Indus River, in what is now Pakistan. In 305 B.C., the Indian prince Chandragupta Maurya regained these provinces from Seleucus, Alexander's successor in Persia, and went on to create the first Indian empire. Under Asoka, Chandragupta's grandson, the

Maurayan Empire reached its height, having brought all but the southernmost regions of the subcontinent under imperial rule. During Asoka's reign (272–235 B.C.), diplomatic and trade relations were maintained with the Greek successor states in Persia, Egypt, and Greece. Edicts by Asoka found in western Pakistan and eastern Afghanistan are in Greek, attesting to the presence of a Greek reading population within the empire itself.

Bactria, a country in what is now Afghanistan and southern Russia, became an independent Greek kingdom in 250 B.C. As the Maurayan Empire declined toward the end of the third century, Bactria extended its rule into northern India (now Pakistan), establishing Indo-Greek principalities there. Greek influence remained strong throughout much of Central Asia and northwestern India until tribal peoples, particularly the Yueh-chih, overran Bactria in around 100 B.C.

Thus for more than 200 years after the death of Alexander, India was in intimate contact with the Hellenistic world. These contacts were of the utmost importance, with the Greek ideal of beauty influencing Buddhist art and the Indian ideals of asceticism and piety influencing western Asian religion. Indian scholars undoubtedly went to Alexandria to translate Indian texts into Greek, eventually returning to India with knowledge of Greek astronomy and mathematics. This was just at the time that Hipparchus was developing his planetary model and the trigonometry it required.

Trigonometry

In my day, trigonometry was the dreary business of looking up sines and cosines in tables and using the numbers so found to calculate the distance across a river. Students today solve their surveying problems by pushing buttons on a calculator.[1] Hipparchus didn't discover how to use trig tables 2,000 years ago—he discovered how to construct them!

Figure 6.1 shows two right triangles with the same angle θ. The Greeks understood very early that such similar triangles differ only in size; that is, they have the same proportions. The ratio a/h of the side to the hypotenuse of the smaller triangle is equal to the corresponding ratio A/H of the larger triangle. This ratio, which depends on only the angle θ, is called the sine of θ. For example, the sine of 22.5° is 0.38268 . . . , which means that in any right triangle with a 22.5° angle, the side opposite the angle is always 0.38268 times

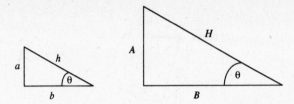

Figure 6.1 Two right-angle triangles with the same angle θ. The two triangles are similar, which means that the ratio a/h is equal to the corresponding ratio A/H. This ratio is called the sine of θ.

the length of the hypotenuse, regardless of the size of the triangle. The interesting question is how Hipparchus found this number.

This digression is critical, because it touches directly on the nature of objective knowledge and whether India ever developed it independently of Greek influence. From elementary Euclidean geometry, it is easy to show that the sine of 30° is exactly 0.5. Hipparchus then proved, using the rigorous methods of Euclidean geometry, a relation that gives the sine of half an angle in terms of the sine of this angle.[2] Thus he could find the sine of 15° from that of 30°, the sine of 7.5° from that of 15°, and the sine of 3.75° from that of 7.5°. He also found a relation that gives the sine of the sum of two angles in terms of the sines of the separate angles. Thus he could find the sine of 22.5° from those of 7.5° and 15°.

The actual calculations are tedious, but by 150 B.C., tables of sines must have existed for angles in increments of 3.75°. These tables were necessary in order to compute the positions of the planets from the Hipparchian model of cycles and epicycles. (For details, see Figure 6.2.) Trigonometry was invented for astronomy, not engineering. Centuries later, Roman engineers made no use of trigonometry in surveying their roads and aqueducts (Mason, 1962).

Hipparchus' work, unfortunately, hasn't come down to us, so we have only indirect evidence as to its contents. We do have a table of sines in quarter-degree steps constructed by Ptolemy three centuries later for the *Almagest,* his great elaboration of the Hipparchian system. The monumental task of calculating sines in quarter-degree steps required him to use a number of advanced theorems probably unknown to Hipparchus. Book I of the *Almagest* is a systematic development of these theorems. In the introduction to this section, Ptolemy wrote:

> We will first show how we can establish a systematic and
> speedy method of obtaining [sines] based on the uniform use of

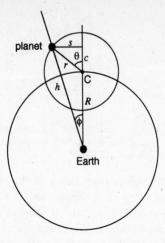

Figure 6.2 In the Hipparchian model, trigonometry is required to calculate the angle ϕ of the planet as seen from the earth in terms of the angle θ and the radius r of the epicycle.

> the smallest possible number of propositions, so that we may not only have the [sines] set out correctly, but may be in possession of a ready proof of our method of obtaining them based on geometrical considerations. (*Almagest,* bk. I; quoted in Heath, 1921, pp. 276–77)[3]

The implication of this passage is that longer treatises on the subject already existed and that Ptolemy was condensing them as much as possible. This statement also shows that the Greeks clearly understood that a table of sines without a demonstration of its validity was worthless.

Hindu Astronomy

Observational astronomy developed independently in India around 2000 B.C. Based on the daily motion of the moon through the zodiac, it is distinct from the astronomy of neighboring civilizations. But by the early centuries of the Christian era, foreign elements had clearly been incorporated into an essentially Indian framework. They used, for example, the Babylonian–Greek division of the circle into 360 degrees, and of the degree into 60 minutes, even though their own number system was decimal.

The astronomical works of the ancient Hindus are called *Siddhantas,* or "established conclusions." These are the works that al-

Mansur had translated into Arabic (see Chapter 5). The best known of these is the *Siddhanta Surya,* which the Hindus believed to have been inspired by the sun. It is a compilation of rules and recipes in words so brief and exact as to be almost unintelligible. For example, the rule for finding the angle ϕ of a planet (Figure 6.2) given its angle θ along the epicycle is as follows:

> Find the . . . Koti-Phala $[c = r\cos\theta]$. . . . It is to be added to
> the radius $[R]$, when the Kendra $[\theta]$ is less than three signs
> $[90°]$ or greater than nine signs $[270°]$; but when the Kendra is
> greater than three signs and less than nine, then the . . . Koti-
> Phala is to be subtracted from the radius.
> Add the square of this result to that of . . . the . . .
> Bhuja-Phala $[s = r\sin\theta]$; the square [root] of the sum is the
> Sighra-Karna $[hyp]$. . . . [Divide s by hyp.] Find the arc whose
> sine is equal to the quotient; the number of minutes contained
> in the arc is called the Sighra-Phala $[\phi]$. (Brennand, 1896, p.
> 220)[4]

The epicyclic model in the *Siddhanta Surya* is much simpler than Ptolemy's and supports the hypothesis that the Indians learned the original system of Hipparchus when they had contact with the West. Greek knowledge was absorbed, however, without the Greek method. That is, the *Siddhanta Surya* is considered a divine work, with the authority for its rules resting on revelation, not reason.

This is nowhere more strikingly revealed than in the table of sines in the *Siddhanta Surya* (Brennand, 1896). This table correctly gives the sines for angles from zero to 90° in steps of 3.75°, indicating that it was originally constructed using Hipparchus' simpler theorems. Remarkably, however, the *Siddhanta Surya* itself gives a rule for constructing the table that is mathematically preposterous. More remarkable still, this totally invalid rule does succeed in generating numbers that are within 2 percent of the numbers in the table, but only for the particular interval of 3.75° and for the particular unit in which the sines are expressed. The same method, applied to any other interval, gives totally incorrect values (see Appendix A). There can be no other conclusion than that the rule was invented to fit a table whose origin had been forgotten.[5]

The table must be of Greek origin, though written in the Indian number system and in Indian units. It was probably calculated around 100 B.C. by an Indian mathematician familiar with the work of Hipparchus.

Despite the relatively primitive state of the Greek science in the

Siddhanta Surya, by stimulating Arabic science, this work played an important role in the history of science. That India didn't itself develop a scientific tradition can be attributed to the Hindu predilection for sanctifying the secular, in contrast with the Greek tendency to secularize the sacred. Furthermore, since all learning was controlled by a religious caste obligated since time immemorial to uphold orthodoxy, the society simply had no place for the skepticism and open speculation required of science.

Bureaucratic China

Traditional China and India, the two great civilizations of eastern Asia, were as different from each other as ancient Greece was from ancient Israel (Coates, 1972). China was basically secular and bureaucratic, whereas India was religious and autocratic. For much of its history, China was more unified, organized, cultivated, and technologically advanced than Europe. The Chinese have navigated by magnetic compass since the twelfth century, 200 years before the Europeans did; they have used the sternpost rudder since the eighth century, 300 years before the Europeans did; they have built their ships with separate watertight compartments to protect against sinking since the fifth century, 1,200 years before the Europeans did; and they have put battens in their sails to increase their efficiency since the first century, 1,800 years before the Europeans got around to it (Singer et al., 1956).

Just as the founding spirit of Greek civilization was the poet Homer, that of China was the scholar Confucius. Devoted to the classics, he is credited with preserving, and perhaps editing, a number of ancient works, such as the Book of Odes *(Shi Jing),* a collection of 305 lyrical poems, and the Book of History *(Shu Jing),* an account of the earliest dynasties of China. Born in 551 B.C., ninety years before Socrates, Confucius promoted scholarship over baseless speculation or mystical insight: "Sometimes I have passed a whole day without eating and a whole night without sleep, giving myself in thought. But it was of no avail. It is better to learn" *(Analects,* xv/30; in Chai and Chai, 1965, p. 44).

Confucianism is preeminently a philosophy of human relations. It emphasizes the importance of humanity *(ren),* righteousness *(yi),* and protocol *(li)* for maintaining harmonious relations between parents and children, husband and wife, the government and the people. Filial love and piety were the foundation and model for all other relations. Government is to the people as parents are to their children. Confucius advised: "Approach the people with dignity and they will be respectful. Show filial piety and kindness, and they will be loyal. Promote those who are worthy, and train those who are incompetent; and they will be encouraged to cultivate virtue" (*Analects*, II/20; in Chai and Chai, 1965, p. 53).

Confucius spent most of his life traveling from state to state in search of a government position. Although greatly respected in his time and often sought out by local rulers, he remained a teacher, never obtaining the high office that would have enabled him to put his ideas of government into practice. This isn't surprising, considering what these ideas were:

> The Master said: "The essentials [of good government] are sufficient food, sufficient arms, and the confidence of the people."
>
> "But," asked Tzu Kung, "if you have to part with one of the three, which would you give up?"
>
> "Arms," said the Master.
>
> "But suppose," said Tzu Kung, "one of the remaining two has to be relinquished, which would it be?"
>
> "Food," said the Master. "From time immemorial, death has been the lot of all men, but a people without confidence is lost indeed." (*Analects*, XII/7; in Chai and Chai, 1965, pp. 53–54)

Free from superstition and mysticism, devoted to peace and respect for the individual, and committed to the education of all, Confucianism seemed to provide the ideal foundation on which to build science. Yet as the extensive research of Joseph Needham and his colleagues has shown, although China made many important technological discoveries critical to the ultimate development of science, it didn't itself develop true science in the sense we are using the term. Although China certainly moved beyond the narrow egocentrism of Israel and India, it didn't reach objectivity. Neither subjective nor objective, Chinese thinking could be called affiliative.

Just as the egocentric mind does not differentiate internal thoughts from external reality, the affiliative mind does not differentiate personal concerns from group concerns. In traditional China, each Chinese saw himself or herself linked to a great network of human and extrahuman relations: ancestors, family, clan, state. Even today, Chinese students, whether American or Asian, are far more motivated by their love for their parents and by the absolute devotion of their parents to them than are European students. This affiliation sustains students when their individual will weakens.

Confucius lived in the middle of China's feudal period (ca. 800–255 B.C.) when there were many small states, each ruled by a local prince and governed by his appointed ministers. He became the teacher of young men ambitious for public office, inculcating them with the notions of *ren, yi,* and *li.* After he died in 479 B.C., Confucianism became one of many schools whose students competed with one another for government positions. It was greatly advanced by the eloquent teachings and writings of Mencius (372–289 B.C.). But after the Qin dynasty united China in 221 B.C., all works of philosophy were ordered burned, weakening Confucianism and obliterating most other schools of thought.

Philosophy and learning revived after the establishment of the great Han dynasty (205 B.C.–A.D. 220). Confucianism was institutionalized by Emperor Wu (140–87 B.C.), who established the mandarinate, the Confucian system of scholar-administrators that was to govern China for the next 2,000 years. Schooled in the Confucian classics, the mandarins were selected by competitive examination to fill administrative posts around the country. It was a nonhereditary system; promising boys from poor families were given scholarships to government schools. Thus although the son of a high official had a cultural advantage over poorer rivals, entry to the system was open to new talent.

It is important to realize that Confucianism didn't create bureaucratic China. Rather, the bureaucracy selected Confucianism from among many other schools of thought as being most congenial to its needs. It is a small shift indeed from Confucius' doctrine of social status—"Let the ruler be ruler, the minister be minister; let the father be father, and the son son"—to the authoritarian notion that it is the right of the ruler to rule and the duty of the inferior to obey. In time, the ethical aspects of Confucianism became so subordinate to its bureaucratic ideology that a twelfth century Confucian could write: "The way of Confucius had not been put into practice for a single day" since the establishment of imperial rule 1,400 years ago (Chai and Chai, 1965, pp. 17–18).

The mandarinate is the key to understanding how China's great technological innovations never affected its ideology. A Confucian education involved the study of works that were ancient when Confucius first collected them. Much of what passed for history was legend, and what passed for deep literature was country doggerel.

Here is an example of one of the 305 Odes that every Confucian scholar had to memorize:

> If you will love me dear, my lord,
> I'll pick up my skirts and cross the ford,
> But if from your heart you turn me out—
> Well, you're not the only man about,
> You silly, silly, silliest lout! (Giles, 1901, p. 14)

Delightful as it is, it was never allowed to be taken at face value. Because the Odes were part of a curriculum that led, through the examination system, to the highest offices in the land, each one was interpreted past recognition. The preceding Ode, for instance, was thought to express the wish of the people of a certain small state that some great state would intervene and end an existing feud in the ruling family. Thus did a rational and democratic system for the selection of government officials degenerate into mind-numbing pedantry.

Nevertheless, compared with the way that officials were selected in other countries—before its revolution, France *sold* its public offices—the system was fair and rational. For almost 2,000 years, China had one of the best governments in the world. It built an extensive system of canals and irrigation works; it operated research and educational establishments; and it ran mining and manufacturing enterprises. In fact, the mandarinate had such prestige accruing from its rigorous education and its monopoly on power that even the sons of wealthy merchants wanted to join it. The most talented young men were continually drained from the merchant class into the mandarinate, strengthening the one at the expense of the other (Boorstin, 1985).

The mandarinate maintained astronomers, mathematicians, and artisans in a variety of government institutions that over the centuries were responsible for many important inventions. Perhaps the most important—paper—was invented by the Imperial Workshops in A.D. 105. It isn't surprising, therefore, that this invention, which was to play such a pivotal role in the advancement of learning when it finally reached Europe 1,000 years later, remained primarily an instrument of the bureaucracy in China.

During the Tang dynasty (618–907), the Office of Mathematics had one professor, one assistant professor, and thirty graduate students who administered the state examinations for the degree of Master of Mathematics. A typical question was "Suppose that there is a field of 15 paces in width and 16 paces in breadth; what is the area?" (Nakayama, 1969, p. 16). Clearly, this wasn't mathematics as we use the term, but simply the training of minor functionaries. The bureaucracy had no need for abstract mathematics, so none was created.

At a higher level were the calendrical mathematicians, responsible for preparing the calendar for the following year. They were from elite families, had passed the highest state examinations in Confucian studies, and were attached to the Office of the Grand Astrologer. This office was of the greatest importance in China, and its work was top secret. As late as the Qing (Manchu) dynasty (1644–1911), the private study of astrology was forbidden.

Although most ancient civilizations maintained state control of calendar making and some aspects of astrology, the Chinese monopoly on astrology was based on the unique concept that the emperor was chosen by heaven and ruled according to the celestial order. Any unusual natural event—whether supernovae, earthquakes, or hailstorms—was a sign from heaven. As one Han emperor explained in an edict: "Our experience in governing has been [only] for a brief time, so that [we] have not been correct in [our] acts, hence . . . there was an eclipse of the sun and an earthquake. We are greatly dismayed" (Nakayama, 1969, p. 45).

Night by night, the court astrologer monitored the motion of the planets to learn what heaven intended, rather than to learn how heaven worked. Similarly, since the second century A.D., the Chinese had had an earthquake detector—a large brass bowl with strikers balanced on its rim—not to provide fast relief to a stricken area, but to monitor what heaven thought of the emperor.[6] That is why the work was so secret. Chinese astrology, like Greek divination, was based on omens, hints from heaven. Properly interpreted and exorcised, disaster might be avoided. This, rather than scientific curiosity, accounts for the meticulous observations of natural events that the Chinese recorded for over 2,000 years.

By the eighth century, Arab traders had reached as far east as Xian, where the great mosque they built still stands. From them, the Chinese acquired some knowledge of Greek astronomy, such as the use of the armillary sphere to plot the motion of the sun and stars. Nevertheless, Chinese astrologers continued to believe that the earth was flat. China, or the Middle Kingdom, as it called itself, was at the center of this flat earth. The emperor's right to rule was bestowed by heaven. The constellations were associated with twelve ancient feudal states. The whole cosmology was centered on the Chinese. This ethnocentric view so permeated their thinking that even when they sailed forth into the outside world, they failed to discover it.

The prime minister was appointed by the emperor from among the top officials in the bureaucracy. Often he wasn't from an aristocratic family and so could speak with the emperor in only the most formal circumstances. On the other hand, the eunuchs who looked after the emperor's wives and children were on intimate terms with him and his family. For this reason, the palace eunuchs at times were able to challenge the mandarinate's monopoly on power.

And so it was that over the objections of the mandarinate, the largest naval expeditions the world had ever seen were undertaken by the grand eunuch Zheng He (Cheng Ho). From 1405 to 1433, he made seven voyages throughout the South China Sea and the Indian Ocean. On his seventh and last expedition in 1431, he sailed for two years with over 20,000 men in hundreds of ships, establishing relations with rulers in Indonesia, India, Arabia, and eastern Africa. His treasure ship was a gargantuan vessel more than 400 feet long, perhaps the largest wooden ship ever built.

Zheng He's seven expeditions were not undertaken for conquest, war, discovery, or any other motive that would justify the incredible cost to a Western mind. Rather, the Chinese believed themselves to be so superior to the rest of the world that they wanted nothing from other peoples except their acknowledgment of this superiority. When Zheng He sailed into a foreign port with his immense flotilla, he brought presents of gold, silver, and gold-embroidered silk as demonstrations of this superiority. To fulfill the tributary obligations demanded of him, the local sultan or monarch had only to admire the generosity and attainments of the Chinese and to reciprocate with local gifts for the Chinese emperor. Because the Chinese gifts given to foreign sovereigns were of greater economic value than the gifts received in tribute, foreign potentates were only too happy to become Chinese tributaries (Boorstin, 1985).

Despite the many strange lands they visited, the Chinese, seeing everything from their own perspective, learned nothing about the outside world. This is wonderfully illustrated by the story that Daniel Boorstin tells about the reaction of the Chinese to the first giraffe brought back from one of these expeditions. Instead of marveling at this incredible beast and being curious about it, the Chinese simply took it to be a *qi-lin,* a mythical unicornlike beast whose appearance was expected whenever China had an emperor of exceptional wisdom and virtue.

The unrivaled power of the emperor allowed him to expend the wealth of China in proclaiming his magnificence. But like building the Great Pyramids of Giza and sending men to the moon, Zheng He's expeditions were too expensive to be sustained by succeeding administrations. The mandarinate had good reason to oppose the expense of the expeditions and the disastrous system of bribing countries to become tributaries. It was finally able to prevent further expeditions on the reasonable grounds that the money was better spent on granaries and irrigation projects.

But it went further than that. In 1433, the year that Zheng He returned from his last expedition, edicts forbidding private maritime trade started to be issued. These purposely undercut the very trading opportunities that the voyages had developed. Over time, harsher and harsher penalties, including death, were levied against building and sailing ships with more than two masts. Although first aimed at foreign trade, they gradually came to encompass coastal trade and fishing as well. Shipbuilding declined, and within forty years the Chinese navy was reduced to a third of its former size (Boorstin, 1985).

The consequence was that when Vasco da Gama, with 170 men and four small caravels, reached India in 1498, there was no Chinese navy to challenge them. The subsequent Portuguese colonization of the South Seas was the result of eighty years of systematic exploration, great courage, and ruthlessness; incredible good luck;[7] and the peculiarities of imperial and bureaucratic China. The Chinese ban on coastal trade resulted in widespread smuggling and corruption, which the Portuguese knew how to exploit when they reached China in 1514.

Why did China, on the verge of becoming a world power, draw back? And why, having drawn back, did it deliberately destroy its own navy, the greatest in the world? The answer to the second question lies in the peculiar ethnocentric and affiliative thinking of the Chinese, which saw the world as a network of interpersonal relationships between parents and children, supervisors and workers, emperor and people. This gave great unity and stability to China because the maintenance of harmonious relations was central to the Confucian doctrine, but it prevented the Chinese from seeing anything outside that network. Nothing of real importance could happen outside China, and certainly there was no danger of invasion from abroad.

Overseas trade was deliberately suppressed to curtail the rising power of the mercantile class. The mandarinate knew that by fostering friendly foreign contacts, Zheng He's expeditions would eventually lead to trade advantageous to China. But such trade would also strengthen the merchants and manufacturers. They would make large fortunes sailing private ships to foreign countries and would in time become an independent force, outside the control of the bureaucracy.

The bureaucracy that brought so much stability to China also foreclosed its future, as it suppressed the very elements—the entrepreneurs and adventurers—who have been the instruments of change in the West. The mandarinate probably didn't cling to power any more tenaciously than any other bureaucracy; rather, like its counterpart in present-day China, it had such total control that nothing could develop beside it. It could encourage new technologies and then stifle their full implementation. For example, China invented printing 600 years before Europe. But in China, printing was just another instrument of the bureaucracy, used to print the Confucian classics and religious texts, whereas in Europe it was, from the beginning, a business catering directly to the public.

The state bureaucracy of classical China was no better than those of communist China and Russia in fully exploiting the creative genius of its people. All severely restricted entrepreneurs to prevent them from becoming rival sources of power. Bureaucracies are effective in establishing a high degree of social order, but when they have too much power, they inevitably become corrupt, inefficient, and self-protective.

Both imperial and bureaucratic interests were served by a perpetuation of China's isolation and ethnocentrism. Without an independent merchant class, rudimentary democratic institutions, and an influx of foreign ideas, there was nothing to oppose China's affiliative thinking. As brilliant as the Chinese were, they didn't develop objectivity, let alone abstract deductive thinking. Classical China's 3,000-year history dramatically demonstrates that science isn't an inevitable outcome from a literate, technologically sophisticated civilization.

Almost all our knowledge of China's technological brilliance comes from the work of Joseph Needham and his Chinese colleagues. This work is often cited as evidence that science did have an independent development outside the Greek mainstream. But Needham himself has always been careful to distinguish ancient and medieval science from modern science, for he was keenly aware of the unique genesis of modern science and the problem that this uniqueness poses:

> The study of great civilizations in which *modern* science and technology did not spontaneously develop obviously tends to raise the causal problem of how modern science did come into being at the European end of the Old World, and it does so in acute form. Indeed, the more brilliant the achievements of the ancient and medieval Asian civilizations turn out to have been the more discomforting the problem becomes.
>
> . . . If you reject the validity or even relevance of socio-logical accounts of the "scientific revolution" . . . you are driven back upon an inescapable dilemma. One of its horns is called pure chance, the other is racialism however disguised. (1969, p. 214)

Needham brilliantly delineated the sociological factors that distinguished China from Europe. Some of these we discussed here, and others are considered in Chapter 7. But showing what factors led to science does not guarantee that such factors must occur in history or that, if they do occur, they must necessarily result in science. Thus, although I fully accept Needham's approach, I don't believe that it removes the element of chance. After all, Europe was Europe by chance, just as China was China. And it is the culture of medieval Europe, not medieval China, that appears decidedly odd in comparison with the other cultures of the world.

It is to medieval and Renaissance Europe that we next turn to look for the critical factors that initiated science. Why did deeply religious Europe, not secular China, discover science? Did science develop because of Christianity or in spite of it? How unusual were the European conditions in comparison with those in the rest of the world? In other words, had Europe not been Europe, how likely is it that science would ever have developed?

7

Towns and Gowns

In *The Divine Comedy,* Dante describes the fourteenth century's concept of the universe. According to Dante, the earth is a stationary globe at the center of a system of nine rotating concentric crystalline spheres. The sun, moon, and each of the five then-known planets is attached to its own sphere. All the fixed stars are attached to the eighth sphere, which encircles the first seven. Beyond the sphere of the fixed stars is the ninth sphere, the *primum mobile,* that carries no stars or planets but that rotates about the earth once every twenty-four hours, carrying the eight inner spheres with it. Beyond the *primum mobile* is the final sphere, the stationary Empyrean Heaven, the abode of God. In this universe, humankind has a unique physical and spiritual place, above the beasts and below the angels (Butterfield, 1965).

Dante's universe is a poetic simplification of Aristotle's geocentric system. By the second century B.C., Hipparchus had employed epicycles to explain retrograde motion, and Ptolemy's system went even further by displacing the earth somewhat from the center of rotation of the planetary orbits. Nevertheless, the medieval mind derived great satisfaction from believing that Dante's religiously conceived cosmography was in accord with Greek science.

The scientific revolution in the seventeenth century destroyed these spheres and this comfortable cosmography, displacing humankind from the center of the cosmos. It was truly a matter of gaining knowledge at the cost of losing paradise. We can get some feel for the tenor of the times from the words of one of Shakespeare's credulous old courtiers: "They say miracles are past; and we have our philosophical persons, to make modern and familiar, things supernatural and causeless. Hence is it that we make trifles of terrors, ensconcing ourselves into seeming knowledge, when we should submit ourselves to an unknown fear" (*All's Well That Ends Well*, II.iii.1–6). How backward Europe made "modern and familiar, things supernatural and causeless" is the subject of this chapter.

Medieval Europe

For 600 years after the fall of Rome, Europe was poor and ignorant. Its illiterate princes ate with their hands in cold castles while the emirs of the East discussed the works of Plato and Aristotle in sumptuous luxury. As the western province of the Roman Empire, Europe had no illusion about its central position in world affairs, and the particular importance that Christians placed on distant Jerusalem—long under Islamic control—kept them free of excessive parochialism. In 1099, the First Crusade succeeded in capturing Jerusalem from the Arabs, thereby establishing the Latin kingdom of Jerusalem. Although Jerusalem was recaptured by Saladin in 1187, European crusaders continued fighting in the Levant until the last Christian stronghold fell in 1291.

From these years of foreign war, Europe was stimulated by the discovery of new ways of thinking and living. Contact with the Arab countries opened up new trading opportunities, which were especially advantageous to Venice and other maritime city-states in a position to do business with the East. More important still was the introduction into Europe of the works of the ancient Greeks, starting in the twelfth century with the first translations into Latin of Arabic editions of Euclid and Ptolemy.

The Crusades were made possible by the slow but steady increase in European prosperity that started in the eleventh and twelfth centuries. Historians have attributed this economic advancement to changes in agricultural technique and technology. I can still remember my medieval-history professor becoming positively lyrical about the importance of the transformation from two-crop rotation to three-crop rotation. His course followed my semester of Greek and Roman studies, and still bedazzled by the brilliance of Greece, I had difficulty concentrating on the plight of the dreary Europeans and their even drearier agricultural problems. It was hard to convince me that anything interesting could have happened in the twelfth century.

But indeed, many things had. One was that shift in crop rotation. In the early Middle Ages, half the fields were planted in any one year, and the other half was left fallow. The fields were reversed the next year. Thus only one-half of the arable land was cultivated at any one time. By the twelfth century, however, a three-crop rotation was being used, in which a field would be planted in winter wheat one year, in a summer grain the next, and left fallow the third. In this way, two-thirds of the land was cultivated, or one-third again as much as with the two-crop rotation (Strayer and Munro, 1942).

More important, the Europeans were able to cultivate this increased acreage without increased labor because of the invention of the horse collar. Incredibly, since ancient times the Europeans—including now the Greeks and Romans—had been harnessing their horses incorrectly. The Chinese had done a better job of it, and Needham (1969) suggests that the horse collar may have come from China. Whatever the case, the Europeans gradually replaced the throat-and-girth harness with a padded collar. The throat-and-girth harness puts pressure on the horse's throat and chest, restricting its breathing and collapsing the veins that drain the blood from the head (Singer et al., 1956). This, needless to say, greatly reduces the load that a horse can pull. But with a padded horse collar, the load is applied to the horse's shoulders, allowing it to work much more efficiently.

Properly harnessed, a horse is as strong as an ox but 50 percent faster. It also has more endurance and so can work several more hours a day. Thus with a horse, a man could plow nearly twice as much land as he could with an ox. Furthermore, the greater speed of the horse revolutionized transportation. Markets grew in size, as farm produce could be economically hauled longer distances. Towns grew up around the larger markets, providing better services to travelers and residents. When a town was large enough, it could obtain that most important of European inventions: a charter.

A charter was a grant by the king of some degree of autonomy to a town or another corporate institution. A town might, for instance, receive the right to make its own criminal laws, administer justice, regulate trade, impose taxes, and contract public works. The merchants in a town were organized into associations, called guilds, that might themselves be chartered to regulate their own business: to set wages, prices, terms of apprenticeships, and qualities of goods. Thus the medieval town had a degree of self-government that hadn't been seen since ancient Greece. In northern and central Italy, the cities became independent states, and some, like Venice, Florence, and Genoa, were actual republics.

Europe suffered a period of greatly diminished freedom in the seventeenth and eighteenth centuries, with the rise of absolute monarchies. But in hindsight, this period was anomalous, ending in civil war in England and revolution in France. The norm was for a high degree of decentralization and local autonomy, in comparison, say, with the centrally administered Chinese and Arab empires. This predilection for independence and self-organization was perhaps the principal characteristic that distinguished Europe from all other civilizations, save ancient Greece itself.

With increased prosperity and freedom, medieval Europe undertook a number of ventures: the Crusades, the building of great cathedrals, and the founding of universities. In the universities, students and scholars studied the works of the ancient Greeks as they became known in Latin translations, first from the Arabic and later directly from Greek sources. Schools of medicine were established to study Hippocrates and Galen, schools of astronomy to study Euclid and Ptolemy, and schools of theology to study Aristotle.

The University of Paris, founded in the twelfth century, became the center of Aristotelian study. There the theologian Thomas Aquinas (1225–1274) studied Latin translations of Arabic translations of Aristotle, a Greek philosopher who had died three centuries before Jesus, in order to demonstrate that the beliefs of the Christian faith were in harmony with reason and logic. Despite their admiration for the clarity of Greek thought, the medieval theologians had no difficulty transforming it to their own purposes. From Aristotle's work, they extracted a complete cosmography compatible with Christian belief. Dante's universe, as described earlier, was part of this cosmography. So, too, was Aristotle's physics, which, because it seemed to prove logically the Christian view of the universe, became linked to orthodox Christian belief. For this, Aquinas was sainted.

Aristotle

Aristotle was born in Stagira in 384 B.C. He studied under Plato at the Academy in Athens, tutored Alexander the Great in Macedonia for a few years (342–339 B.C.), and then founded the Lyceum, his school in Athens where for twelve years (335–323 B.C.) he lectured on methods of inquiry, physics, metaphysics, and art. His extant work consists largely of these lectures as recorded and edited by his students and successors at the Lyceum (Bowen, 1972).

Aristotle's contribution to science has been widely misunderstood. He has been hailed by some as the greatest scientist of antiquity and decried by others as the greatest impediment to the development of modern science. The truth is that Aristotle, whose death coincided with that of his pupil Alexander, lived near the beginning of the Golden Age of Hellenic science. Two of the greatest scientists of antiquity, Archimedes and Ptolemy, came 100 and 500 years after him. Greek science, especially astronomy, advanced well beyond Aristotle in the ensuing years.

This isn't to say that Aristotle wasn't a mature scientist. By his time, the level of scientific thinking was as good as it gets. Aristotle's astronomy was based on the work of Eudoxus (fl. 350 B.C.), arguably the greatest mathematician of antiquity. Eudoxus is credited with the theory of proportions that Euclid presents in Book v of the *Elements*. His sublimely subtle definition of proportions[1] is essentially the same as the definition of irrationals used in the modern theory of numbers (Heath, 1908, 1913; Landau, 1928/1951).

Eudoxus is credited also with the first geometric planetary model. But his work has been lost, and we have only a few paragraphs about his system from Aristotle and Simplicius, a sixth-century commentator. From these, scholars have reconstructed what they believe to be an accurate description of his remarkable achievement. It is of interest to us because it shows that Aristotle was working within a tradition of great sophistication.

In the system of Eudoxus, the motion of each planet is described by the combined motion of four spheres of nearly equal diameters rotating about the earth. The planet itself sits on the inner, or fourth, sphere. The first sphere rotates once a day, producing the daily motion of the planet. The axis of the second sphere is fixed to the first. It rotates with the synodic period—the time between successive periods of retrograde motion—and so describes the planet's average seasonal motion against the background of stars. The axis of the third sphere is fixed to the second, and the axis of the fourth is fixed to the third. As the third and fourth spheres rotate, the planet, which is a point on the fourth sphere, follows a complex figure-eight trajectory on top of its seasonal motion (Heath, 1913). It is this that produces the retrograde motion. The details aren't important, but it is important to recognize that only a mathematician of the highest caliber could describe the motion resulting from such a set of interconnected spheres, let alone determine the speeds of rotation and the orientations of the spheres' axes that best fit the astronomical facts. A fifth sphere was added in some cases, bringing the number needed to describe the seven celestial bodies to thirty-three (Heath, 1913).

Although Eudoxus probably thought of his spheres as mathematical constructs for calculating the planets' motions, Aristotle took them to be physical. The moon, for example, is physically embedded in its sphere, and as the sphere rotates, the moon naturally keeps its same face toward earth. This explanation is perfectly reasonable, given the facts of Aristotle's time. (Our modern explanation requires the moon to rotate on its axis in exactly the same time as it revolves about the earth, an incredible coincidence that can be understood only on the basis of Newtonian mechanics applied to tidal forces.)

Aristotle went further, however, and assumed that the spheres of all the planets were connected. This forced him to add twenty-two other spheres to undo what the spheres of the outer planets did to the spheres of the inner planets. All in all, the system explained most of the observed features of planetary motion, though it had many failures as well.

For example, the brightness of Venus varies appreciably, indicating that its distance to the earth changes. The concentric sphere model couldn't explain this, since each planet kept a constant distance from earth. It may be thought unscientific to keep a fifty-five-sphere model that is unable to explain such an elementary fact, but the situation is no different today. The standard model in particle physics has forty-five fundamental particles, and it too has its failures. But because it explains so much and is based on such beautiful mathematical principles, its supporters are confident that its shortcomings will someday be resolved. Doubtless the supporters of the concentric-sphere model felt the same way.

Aristotle thought it self-evident that uniform motion about the center of the universe was "natural" and so not in need of any further explanation. He also thought it self-evident that the only other "natural" motions were motion toward the center and away from the center. Heavy matter, solids and liquids, move toward the center, and light matter, gases and heat, move away from it. All earthly matter is composed of the four elements—earth, water, air, and fire—that are the essences of solid, liquid, gas, and heat. The natural abode of earth and water is the center, toward which they rush when separated from it. The center of the earth thus coincides with the center of the universe because all the material of the earth piles up around the center. Heavy objects fall toward the earth because they are trying to reach the center. Since celestial objects, such as the sun, moon, and planets, move neither toward nor away from the center but around it, they cannot be made of any of the four elements. Celestial objects are made of a fifth material, which is perfect and changeless.

In Jean Piaget's last book, he and science historian Rolando Garcia discuss the psychological origin of science. For them, Aristotelian reasoning is based on pseudonecessity, a faulty mode of thought that confuses what is with what has to be (Piaget and Garcia, 1989). Thus, for Aristotle, all natural motion had to be toward, away from, or around the center. A simple body could have only one of these motions, and so on. Piaget had observed a similar unwarranted restriction on possibilities in children asked to solve certain open-ended questions. He believed that Aristotle may not have passed through this stage of intellectual development. According to this view, modern science developed by going beyond pseudonecessity, just as children ultimately do.

As do Piaget and Garcia, most scholars of the scientific revolution believe that a great qualitative change in thinking took place between Aristotle and Galileo. I, however, believe that the major change in human thinking came with the Greeks' discovery of objectivity. In my view, Aristotle's thinking isn't really any worse than that of a modern scientist. Certainly his speculations went well beyond what could be justified by the knowledge of his time. He made too many "educated" guesses, and that isn't science, strictly speaking. But it's well within the tradition of both ancient and modern science. Aristotle's students may have understood this perfectly well, but were as interested in his speculations about the universe as we are in Stephen Hawking's (1988). It isn't Aristotle's fault that medieval theologians elevated his speculations to religious dogma.

Aristotle's universe was spherical. The earth coincided with the center because earth and water naturally moved toward this center. The only other natural motion was the motion of air and fire away from the center and the motion of celestial bodies around the center. Violent, or unnatural, motion requires the continuous action of a mover, what we would call a force and what Aristotle's translators call a movent. Push a book across a table with your hand. While your hand is in contact with the book, the book moves. As soon as contact is broken, the book stops. Thus Aristotelian physics, with its belief in the need of a force to sustain motion, is well founded on everyday experience.

It does have its problems, however. An arrow shot from a bow is propelled forward by the action of the bow string. What maintains its motion after it leaves the bow? According to Aristotle, it's the air. The moving arrow causes some air to move, which rushes back to cause the arrow to move; the moving arrow then causes more air to move; and so on. The air and the arrow alternate as the movent and the moved. Why, then, does the motion ever stop?

> The motion begins to cease when the motive force produced in one member [the air or the arrow] no longer causes the next member to be a movent [cause of motion] but only causes it to be in motion. The motion of these last two—of the one as movent and of the other as moved—must cease simultaneously, and with this the whole motion ceases. Now the things in which this motion is produced are things that admit of being sometimes in motion and sometimes at rest, and the motion is not continuous but only appears so. (Aristotle, *Physica,*VIII; quoted in Piaget and Garcia, 1989, p. 40)

According to Aristotle, all motion is in either a straight line or a circle. The arrow moves in a straight line horizontally until its motion ceases and then it falls straight to the ground. The arrow's horizontal motion is discontinuous, an alteration of stops and starts, and can occur only in air or water. A vacuum can't exist because there objects would move infinity fast, which is impossible.

To anyone familiar with Newtonian physics, all this is nonsense. But to anyone unfamiliar with Newton, which was everyone in the Middle Ages, it's no less reasonable than any other bit of scientific mumbo jumbo. Consider this bit of Newtonia:

> A block sitting on a table is at rest, and so again, by Newton's first law, there must be another force acting on it in addition to the force of gravity. The surface of the table is slightly deformed by the block and, as a consequence, exerts an upward force on the block. This force, which is directed perpendicular to the surface that produces it, is called the normal force. Since the sum of the force of gravity on the block and the normal force on the block is zero by Newton's first law, it follows that [the normal force] must be directed opposite to [the force of gravity]. (Cromer, 1980, pp. 31–32)

In Newtonian mechanics, it takes two forces to keep a block at rest on a table. One is the invisible force of gravity, the weight of the block. The other is the force of the table on the block, which, to the uninitiated, is as arbitrary as Aristotle's moving air.[2]

According to the historian Herbert Butterfield,

> The supreme paradox of the scientific revolution is the fact that things we find it easy to instill into boys at school, because we see that they start off on the right foot—things which would strike us as the ordinary natural way of looking at the universe, the obvious way of regarding the behavior of falling bodies, for example—defeated the greatest intellects for centuries, defeated Leonardo da Vinci and at the marginal point even Galileo, when their minds were wrestling on the very frontiers of human thought with these very problems. (1965, pp. 1–2)

But as anyone who has either taken a physics course or taught one can attest, there is nothing easy about it. The problem of motion defeated the best minds for centuries, not because they weren't prepared to tackle it, but because it's so difficult.

By the thirteenth century, European scholars had absorbed much of the essence of Greek science. The universities that sprang up all over Europe in the late Middle Ages (ca. 1200–1500) provided employment for teachers and scholars, educational opportunities for tens of thousands of students, and environments in which new ideas and stimulating thoughts were exchanged as students and faculty traveled among them. Roger Bacon (1214–1294) at Oxford was a leading figure in the cause of science, stressing the value of observation, mathematics, experimentation, and hypotheses and deduction. New ideas on motion were advanced. In Paris in the fourteenth century, Jean Buridan and Nicole Oresme, bothered by Aristotle's "explanation" of why an arrow continues to move after it leaves the bow, struggled with the notion of "impetus" or "impetuosity" as something inside a moving body that carried it forward. Yet for all this, no real progress was made in understanding the laws of motion, and science couldn't progress without this understanding.

The principal obstacle to discovering the laws of motion is that we live on the surface of a large revolving planet. The gravitational pull of this planet on all matter on its surface, together with friction and air resistance, conspire to disguise these laws. Because of gravity, it takes two forces, canceling each other out, to keep an object at rest. Physics teachers must expend a great deal of effort devising clever demonstrations—virtual parlor tricks—to illustrate Newton's laws, because these laws are for an infinite universe that has no preferred place or direction. Living, as we do, on a large revolving planet, the universe appears just the opposite. It's "self-evidently" Aristotelian: finite and spherical, with all natural motion oriented toward, away, or around the center.

The problem of motion couldn't be solved until it was realized that we are living on a revolving, orbiting planet and that the center of this planet has no fundamental significance. Only with this shift from an Aristotelian to a Copernican universe could the mind finally free itself from the bonds of geocentrism. It's doubtful whether this shift would have ever been made within the tight confines of medieval scholasticism. But other forces, outside academia, were at play. To understand these, we must leave the gowns for a while and examine what was going on in the towns.

Capitalism

By the fifteenth century, Europe was transforming itself rapidly from a feudalistic society to a capitalistic one. By this I mean that wealth and power were shifting from the hereditary nobles to the merchants and manufacturers. A whole new class was being empowered with an ideology very different from that of the nobility and the clergy. Whereas the nobility was concerned with privilege and ancestry, the capitalists were concerned with risk, enterprise, and the accumulation of wealth. While the nobility was busy passing sumptuary laws—laws forbidding the bourgeoisie from wearing certain materials, such as ermine—the merchants were busy establishing a banking system. While the clergy were busy debating what a Christian rate of interest was, the manufacturers were busy replacing the medieval system of independent artisans with hired factory workers (Tuchman, 1978). The way that the enterprising spirit of the late Middle Ages swept over the established ideologies of the time is being replayed today in Eastern Europe as it makes the transition from communism to capitalism.

In medieval Europe, the entrepreneurial spirit launched the age of discovery. In 1488, after seventy years of exploration along the western coast of Africa, the Portuguese sailed around the Cape of Good Hope. In the next decade, they explored the Indian Ocean, arriving in India in 1498. In the same decade, Christopher Columbus brought the Spanish flag to a new world. These epic adventures, and the flood of stories of strange lands and exotic peoples that followed, reached an ever-growing literate population through the new medium of the printed book.

Before printing, each book had to be hand copied, a time-consuming, expensive proposition that was controlled by long-established guilds. For centuries, what was copied was strictly limited; then suddenly there were thousands of titles available at reasonable price. From the beginning of printing in Europe in about 1455 to 1500—the period known to bibliophiles as the incunabula, or cradle years—over 35,000 titles were published. The sheer magnitude of the industry frustrated the attempts of the authorities to control it.

The Chinese, of course, had had block printing since the eighth century. But its primary purpose was authentication, not dissemination. The authorities wanted to have official editions of all important works. In 953, the Chinese National Academy printed an authoritative edition of the Confucian classics in 130 volumes. Block printing was ideally suited to this purpose, for although each page had to be individually carved in wood or clay, the result was an unchangeable stereotype from which copies could be made whenever needed.

The Chinese also used block printing to make playing cards, and this may have introduced the technology to Europe. By the early fifteenth century, playing cards and religious pictures were being printed in Venice, Strasbourg, and Nuremberg. Johann Gutenberg (1397–1468), though trained as a goldsmith, might have been a partner in one of the Strasbourg printing firms. There he seemed to have conceived of the idea of interchangeable type, and for years he worked in secret on his invention (Boorstin, 1985).

Although simple in concept, the technical problems were challenging. Each letter was formed on the top of a metal stem so that it could be easily handled and yet firmly held in its position in the printing tray. A method had to be devised for quickly casting thousands of types of precisely the same length. Special ink had to be formulated and a procedure devised for making a clear impression on paper. For this, Gutenberg modified the press already in use for binding hand-copied books.

His famous Bible was printed in about 1455, but by then he was in so much debt to his backer, Johann Fust, that Fust won possession of his press, type, and molds. The firm of Fust and Schöffer became the first European publishing house. Other houses sprang up quickly, as the technologies required to cast type and build presses were easily copied in fifteenth-century Europe. William Caxton brought the first press to England in 1477, and by 1500 there were 150 printing firms in Venice alone (Boorstin, 1985).

In Venice, the printer Aldus Manutius introduced the modern octavio-sized book, which greatly reduced the cost of books. He was the first to print books in Greek, to number the pages of a book, and to use a title page to announce the author to the reader. Thus by the beginning of the sixteenth century, inexpensive hand-held books, complete with an index, were becoming commonplace.

The importance of printing—or, rather, of the inexpensive book—to the subsequent history of Europe cannot be exaggerated. Printing alone, as we know from the history of China, isn't enough. China had an educated elite, inventive craftsmen, and even an age of exploration in the fifteenth century. But the mandarinate's tight control over printing prevented it from ever becoming a vehicle for independent expression. In Europe, on the other hand, printing was, from its very beginning, an enterprise undertaken to make money by selling books directly to the general public. This allowed, really for the first time, the direct exchange of ideas among the people of a whole continent. Theologians, philosophers, poets, and pornographers could reach their audiences over the heads (or behind the backs) of the authorities. The writings of Luther and Calvin swept like a fire storm across Europe in the 1520s and 1530s. In a few decades, half of Europe had ceased its obedience to the pope, breaking Rome's monopoly on religious doctrine. Now a book banned by the authorities in one country might find acceptance in another: Copernicus's great work was published in Leipzig after Luther personally objected to its publication in Nuremberg.

Martin Luther was a local vicar and professor of theology at the University of Wittenberg. A pious man, he saw the church being corrupted by the large sums of money flowing to it from its controversial practice of selling indulgences. In 1517, he publicly protested this practice by posting his famous ninety-five theses on the door of the castle church. He was urging restoration of the church's lost piety, not revolution. But matters quickly got out of hand as business and political interests became involved, and his excommunication in 1521 launched the Reformation.

Copernicus

In rough analogy, Nicholas Copernicus, an astronomer and a church official, was a devout Roman Catholic and Aristotelian who saw that the Ptolemaic system wasn't in strict accord with Aristotle's spherical universe. In Aristotle's universe, all the celestial bodies revolve uniformly about the center. But in the Ptolemaic system, they revolve on epicycles, and the centers of these epicycles revolve about points displaced from the center of the earth (see Figure 5.6). Instead of destroying the Aristotelian orthodoxy, Copernicus was trying to restore it (Butterfield, 1965).

The Polish-born Copernicus (1473–1543) studied astronomy in Krakow before studying Greek, mathematics, medicine, and canon law at the Universities of Bologna and Padua. These were exciting times, with new Greek texts appearing every month or so. It was from these that Copernicus learned that not all the ancient Greeks agreed with Aristotle and that some had even suggested that the earth revolved about the sun. Greek thought still dominated European thinking, and would for another 100 years, but now, as more ancient books were published, you could pick your own Greek.

By 1514, Copernicus had privately circulated the ideas of his heliocentric system. In 1533, he even went to Rome to lecture Pope Clement VII on the subject and received papal approval of his work. Thus the Copernican system was generally known before the publication of *De Revolutionibus orbium* in 1543. The first printed copy of this, the most important scientific book since Ptolemy, was placed in the old astronomer's hands just a few hours before he died.

In the Copernican system, the daily motion of the sun and stars across the sky is just an apparent motion due to the rotation of the earth about its axis. In addition, the earth and the other five planets move in concentric circles about a stationary sun. Only the moon still orbits the earth. Each planet takes a fixed time, called its period, to go once around its orbit. For the earth, of course, the period is 365 days, or a year. A simple calculation shows that Mars must have a period of 686 days if the earth is to overtake it every 780 days (Figure 7.1). Remember that 780 days was the time interval between successive periods of retrograde motion. In the Copernican system, retrograde motion is no longer a real motion of Mars, but only an apparent motion that occurs every 780 days as the faster-moving earth overtakes Mars.

For Copernicus, the strongest virtues of his system were the

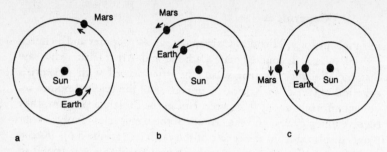

Figure 7.1 The heliocentric system of Copernicus for the earth and Mars: (a) the earth orbits the sun in 365 days, and Mars orbits it in 686 days; (b) half a year later, as the earth overtakes Mars, Mars *appears* to move backward (retrograde motion); (c) 780 days later, the earth overtakes Mars again, but at a different point in its orbit. Compare this with the geocentric system in Figure 5.5.

restoration of true, or nearly true, circular motion and the elevation in importance of the sun, to which he gave great mystical significance. For later generations of astronomers, especially Galileo, its principal virtue was its mathematical beauty. Retrograde motion, which had held such awe for astronomers for thousands of years, suddenly had no more significance than the apparent "motion" of the trees that one passes in a carriage. And the most peculiar features of the Ptolemaic system—that the sun and each planet had a cycle of 365 days as well as a cycle of 24 hours—were eliminated in favor of the single orbital period of 365 days for the earth about the sun and a single rotational period of 24 hours for the earth about its axis. Now, instead of the entire vault of heaven rotating once every 24 hours, only the earth had to move.

For us, the principal virtue of the Copernican system is that it evoked intense discussion. For the first time, Aristotelians were forced to defend their doctrines against bold attack. The earth couldn't be revolving about the sun, they said, because if it were, the apparent positions of a nearby star would shift during a six-month period, as the earth moves halfway around its enormous orbit. Nor could the earth be rotating about its axis once a day because then a stone dropped from the top of a tower would hit the ground more than 1,000 feet west of the tower, since this is the distance that the surface of the earth would have moved in the time the stone fell. For the same reason, if the earth rotated under its atmosphere, 1,000-mile-an-hour winds would be produced.

In response to the first point, Copernicus argued that the stars must be so far away in comparison with the diameter of the earth's orbit that the resulting shifts in their positions are too small to be observed. This meant, given what could be observed with the unaided eye, that the distance to the nearest star was more than 1,000 times the diameter of the orbit. This made the Copernican universe much larger than the Aristotelians were ready to accept.[3]

Galileo

An adequate response to the point about the falling stone couldn't be given until the problem of motion was solved. This was the task that Galileo Galilei (1564–1642) set himself. Because of its simplicity, the Copernican system was being used at the time to make astronomical calculations, though the Ptolemaic system was still generally believed to describe the actual motions of the planets. To Galileo, however, the mathematical economy of the Copernican system won out over its failure to conform to Aristotelian physics. The physics must be wrong.

Galileo was able to succeed where others had failed because as a dedicated Copernican, he was no longer impeded by the idea that the earth was at the center of a spherical universe. His mind was partially free of the geocentrism that our proximity to earth so naturally engenders. Furthermore, unlike his predecessors, Galileo had some idea of what he was looking for: a physics in which the motions of the earth would not be noticeable to the human beings living on its surface.

The problem boiled down to this: Does a stone dropped from the mast of a moving ship hit the deck at the base of the mast or some distance behind the mast? Galileo's famous book *Dialogue Concerning the Two Chief World Systems* addressed this question in the form of a conversation among three Italian gentlemen. At once a scientific treatise and a piece of literature, this truly great work can still be read with interest. In style and language—it was written in Italian, not Latin—*Dialogue* was intended to appeal to a large audience.

The church had officially condemned the Copernican doctrine in 1616, and Galileo was forbidden to advocate it, although he could discuss it. Writing under the stern stare of the Inquisition, he dared not criticize the church's growing restrictions on scientific inquiry. Instead, he tells his readers:

> Complaints were to be heard that advisers [to Rome] who were totally unskilled at astronomical observations ought not to clip the wings of reflective intellects by means of rash prohibitions.
>
> Upon hearing such carping insolence, my zeal could not be contained. . . . I propose in the present work to show to foreign nations that as much is understood of this matter in Italy, and particularly in Rome, as transalpine diligence can ever have imagined.
>
> . . . I hope that from these considerations the world will come to know that if other nations have navigated more, we have not theorized less. (1632/1967, pp. 5–6)

Unfortunately, wit didn't save Galileo, for with the rapid growth of Protestantism, the Inquisition had lost its sense of humor.

Simplicio,[4] the gentleman who takes the Aristotelian position in the *Dialogue,* is a wonderfully satiric character, at times pompous, obsequious, or downright ignorant. I can well imagine that Galileo was prosecuted as much for ridiculing the philosophical and literary pedants of his time as for advocating the Copernican system. Salviati is Galileo's alter ego, and Sagredo represents the interested layman, exactly the person for whom the work was written.

In discussing the stone falling from the mast of a moving ship, Simplicio is certain that it will hit the deck in back of the mast, though he has never performed the experiment because he is certain "that the authorities who adduced it had carefully observed it." Salviati is equally certain that the stone will hit the deck at the foot of the mast, though he too admits to never having done the experiment. Then there ensues a discussion of a ball rolling on a hard polished surface. Since it would gain speed rolling down an inclined surface and would lose speed rolling up such a surface, it must neither gain nor lose speed rolling on a horizontal surface. Returning to the stone falling from the mast of a ship, Galileo argues that it too must maintain the same horizontal speed as the ship, as there is nothing to impede it.

Galileo here is wrestling with the law of inertia: A moving object will continue to move in a straight line at a constant speed unless something (a force) acts on it. It is straight-line motion, not circular motion, that an object "naturally" executes when there is no force on it. We don't detect the motion of the moving earth, just as we don't detect the motion of a moving jetliner, because we are moving along with it. No winds are produced because the atmosphere moves along with the earth.[5]

Galileo's peculiar use of thought experiments (the balls rolling on hard surfaces), rather than real experiments, is deliberate. He had made many observations of the effect of motion on falling bodies, but here he wants to defeat Aristotle with Aristotelian-style arguments. Galileo was heavily influenced by Archimedes, whose writings were first published in Europe only in 1543, the same year that *De Revolutionibus* was published. In fact, with Galileo we may say that Europe had at last caught up with Archimedes in the application of mathematics to physics.

Newton

With Isaac Newton (1642–1727), European science finally came into its own. In his famous *Philosophiae naturalis principia mathematica* (1687), Newton developed the first complete mathematical theory of the cosmos. The word *theory*, as used in the natural sciences, doesn't mean an idea tentatively held for purposes of argument—that we call a hypothesis. Rather, a theory is a set of logically consistent abstract principles that explain a body of concrete facts. It is the logical connections among the principles and the facts that characterize a theory as truth. No one element of a theory, not a single fact or principle, can be changed without creating a logical contradiction that invalidates the entire system. Thus, although it may not be possible to substantiate directly a particular principle in the theory, the principle is validated by the consistency of the entire logical structure.

So it is with the *Principia,* which is written in the style of Euclid, with axioms, postulates, and proofs. And even though today we teach Newtonian physics using the differential calculus that Newton invented, the *Principia* itself used only the classical methods of geometry in proving theorems.

By Newton's time, the basic facts of planetary motion were summarized in three laws, discovered by Johannes Kepler (1571–1630). Kepler had succeeded in deducing the actual figure the planets make in orbiting the sun, by subtracting the earth's motion from the planets' apparent motions as viewed from the earth. This great achievement was possible because Kepler had obtained the superb planetary observations of the Danish astronomer Tycho Brahe (1546–1601), for whom he had at one time worked. Brahe was the first astronomer to keep night-by-night records of the planets' positions. All previous astronomers, including the Babylonians, recorded only the time when a planet was in a special position, such as the start of its retrograde motion.

Kepler was pervaded with mystical notions about numbers. He believed, for instance, that the spacing of the five planets had something to do with the five Platonic solids. Much of his work, therefore, is invalid, but from the mass of it can be extracted three results that correctly describe the planetary orbits. Kepler's first law states that the orbit of each planet is an ellipse, one of the conic sections investigated by Apollonius 1,800 years earlier. The sun is located at a point, called the focus, situated on the long axis of the ellipse. The elliptical orbits of the planets are nearly circular, but the sun is considerably off center. Kepler's second law states precisely how the speed of a planet varies with the planet's position on its ellipse, being greater the closer the planet is to the sun. Kepler's third law states how the time it takes for a planet to move once around its ellipse is related to the distance of the planet from the sun.

Newton knew from the work of Galileo that a planet would move in a straight line at a constant speed if there was no force on it and thus that a force was needed to hold the planets in their orbits about the sun. This is the reverse of Aristotelian physics, which held that motion in a circle at a constant speed was "natural" and therefore required no mechanism to maintain it.

A planet moving at a constant speed in a circle (Figure 7.2) is constantly changing its direction of motion. If there were no force on the planet at point A, it would continue to move in a straight line, but because it does move in a circle, when it reaches point B it has "fallen" the distance *h* toward the center of the circle. Thus for circular motion, the force must be directed toward the center. For

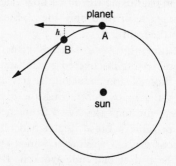

Figure 7.2 In moving from A to B, the planet falls a distance *h* toward the center.

motion in an ellipse, as described by Kepler's first and second laws, Newton showed mathematically that the change in motion is always directed toward the point where the sun is located. Furthermore, he found that the force of attraction between the sun and the planet must decrease as the square of the distance between the sun and the planet if the planets were to obey Kepler's third law. Some of these brilliant deductions were made by Newton while still a student at Cambridge. (Actually, he was home at the time, as the university was temporarily closed because of an epidemic.)

This sun-centered force couldn't be unique to the sun, however, for the moon orbits the earth, and Galileo had discovered with his telescope that similar moons orbit Jupiter. Thus there was a universal force connecting all the astronomical bodies. What was it? The story, of course, is that while pondering the question one day, he saw an apple fall from a tree, giving him the connection he needed (Ferris, 1988). Could the force that the earth exerted on the apple— and on everything on the earth's surface—be the same as this universal force among the astronomical bodies? By this time, science had progressed beyond the point that idle speculation would suffice. A calculation was needed.

The dropped apple falls sixteen feet in the first second. If the force between the earth and the moon is of the same kind as the force between the sun and the planets, it must likewise decrease as the square of the distance. Now the moon was known to be about sixty times farther from the earth's center than the apple, so that in one second the distance h that the moon "falls" should be smaller by a factor of 60 squared, or 0.0044 foot (16 divided by 3600). Newton calculated this distance geometrically from the radius of the moon's orbit and the time for the moon to orbit once around the earth. The result is exactly the 0.0044 foot required if the force on the moon is the same as the gravitational force on earth, reduced by the square of the distance from the earth's center.

Newton did this critical calculation before he was twenty-four. However, because he used some incorrect values for the ratio of the moon's distance to the radius of the earth, he found that the two calculations of h agreed only "pretty nearly" with each other (Jeans, 1951). He therefore didn't publish this work or his demonstration that the inverse square force would result in elliptical orbits until urged to do so by Edmond Halley in 1684. The result was his monumental work, the *Principia,* published in 1687.

Newton showed, with mathematical exactitude, that the everyday force we experience as weight extends into the heavens, acting on the moon with a magnitude diminished by the square of the distance of the moon from the earth. Further, all bodies exert this force in proportion to their mass, with the mass of the sun supplying the force that holds the planets in their orbits. Newton developed a complete theory of motion in general, of planetary motion in particular, and of the force that holds the universe together. Thus he showed that human experience and the human mind had the capacity to penetrate the cosmos.

The impact of Newton on his time is best caught in the unabashed admiration of Alexander Pope's couplet:

> Nature, and Nature's law lay hid in night:
> God said, *Let Newton be!* and all was light.

By Newton's time, there was a conscious acknowledgment that science had entered a revolutionary period. Aristotelianism was dead. New approaches involving the subtle combination of mathematics and experimentation were replacing long chains of speculative reasoning based on "self-evident" truths. The race was on to find the system that would put it all together. Newton won the race in part because, as he put it, he had stood on the shoulders of giants and in part because he just happened to be the biggest giant of them all.

Reasons for the Scientific Revolution

In reviewing the story of the scientific revolution, four points stand out clearly.

First, European science was a direct continuation of Greek science. Indeed, for centuries the medieval scholars dared not have a thought that wasn't written by Aristotle or some other equally worthy Greek.

Second, the distinctive feature of European culture was its tendency to develop autonomous self-governing institutions. The principal institution was the town, which in turn spawned subsidiary institutions such as guilds and universities. These decentralized and diverse institutions offered stability and continuity to Europe's intellectual life despite the political and religious turmoil of the time.

Third, capitalism provided a force and a class that could stand up to the nobility and the clergy. The sons of the bourgeoisie were the students and the professors of the universities.

Fourth, the printing of inexpensive books spread new ideas among the learned and also made possible the education of an increasing number of students from the middle class. An education based solely on lectures was now replaced by one based on reading an ever-growing list of important authors. Further, the market for books encouraged the writing of them. Thus a creative cycle was established that, over the course of the seventeenth century, was to transform the thinking of Europe.

According to the developmental view of thinking discussed in Chapter 2, the modes of thought available at each stage of life depend on the mental structures developed in the earlier stages. It is impossible, therefore, for a person raised with medieval categorical thinking to comprehend modern relational thinking. Rather, the transition from the medieval to the modern had to be gradual, as the occasional insight of one generation was incorporated into the education of the next.

Thus the scientific revolution was dependent on a stable educational system that provided the channel for reshaping the thinking of subsequent generations. The thirteenth-century scholars didn't have the background to understand Euclid or Ptolemy fully, but their students, and their students' students, did. Step by step, the educational system of Europe educated itself.

Europe in the late Middle Ages resembled ancient Greece in several respects. It had a decentralized political system with many autonomous centers of power, a prosperous and independent merchant class, and a stable educational system that had discovered abstract deductive reasoning. In addition, Europe had printing, which greatly enlarged the scope and accelerated the rate of interaction among the learned. But the decisive factor in producing the scientific revolution was the sheer number of scientists who were at work in the seventeenth century.

Attention is usually focused on the new ways of thinking that were emerging at this time. But the number of new ideas depends on the number of thinkers. By the seventeenth century, a critical level of scientific activity had developed that, through synergism, produced a qualitative change in the scientific process.

What was this change? How could a quantitative increase in the number of scientists produce a qualitative change in what they were doing? What were they doing? What is modern science, anyway? What assurance do we have that modern science is ultimately on any sounder basis than was Aristotelian science? Could our science someday be found to be as wrong as Aristotle's was?

These questions are discussed in Chapter 8. We have put them off until now because the answer we propose—though powerful and to the point—is controversial. It can't be given a fair hearing without the background that this, and the preceding chapters, have provided.

8

Science and Nonsense

A concise description of science is surprisingly elusive. Is science to be equated with its products—the cure of disease, the flight of airplanes, the magic of the computer? Or with its methods—mathematical equations and laboratory experiments? Or more subtly with its metaphysics—a belief in objective reality and in cause and effect?

Each of these does capture something of what the word *science* means to most of us, but none is completely satisfying. A description based on material products excludes mathematics and evolution. A description restricted to traditional mathematics and experimentation excludes the historical sciences, such as paleontology, and new methods, such as computer simulations (Yonezawa, 1993). And the metaphysics of nineteenth century science had to give way to the realities of twentieth century physics.

Yet at any time, most scientists agree on what is and what isn't science. Extrasensory perception is not. Parity violation is. Channeling isn't. Cosmology is. Chiropractic and homeopathy aren't. Relativity and quantum mechanics are. What distinguishes science from its imitators and rivals?

Public Knowledge

The British physicist John Ziman (1968) offered a very simple answer that works surprisingly well. Science is the search for a consensus of rational opinion among all competent researchers. The various descriptions of science given earlier are to be understood as manifestations of this goal. The products of science give empirical proofs of its theories. Mathematics and experimentation provide powerful arguments with which to convince and persuade. The metaphysics of science defines a public world about which a consensus is possible.

Science, in this view, is an extension of rhetoric. It was invented in Greece, and only in Greece, because the Greek institution of the public assembly attached great prestige to debating skill. There were teachers of rhetoric and debate as early as Homer and probably much earlier. Over the centuries, the rules of argumentation were developed: how to construct a strong argument and how to counter a weak one. A geometric proof is a logical argument constructed to withstand the strongest counterargument. It's the ultimate rhetorical form.

Geometrical proofs were invented by mathematicians, not rhetoricians, but to a Greek raised in a culture that valued sound argumentation so much, the unsupported recipes of oriental mathematics would have been as unacceptable as the dogmatic preachings of wandering prophets. He would want a convincing argument. In the process of developing such arguments, the early Greek mathematicians discovered formal logic and thereby transformed Eastern numerism into true mathematics.

Science, like democratic politics, is a social activity. In fact, as Ziman emphasized, this is the essential part of its nature:

> The scientific enterprise is corporate. . . . It is never one individual that goes through all the steps in the logico-deductive chain; it is a group of individuals, dividing their labour but continuously and jealously checking each other's contributions. The cliché of scientific prose betrays itself "Hence we arrive at the conclusion that . . ." The audience to which the scientific publications are addressed is not passive; by its cheering or booing, its bouquets and brickbats, it actively controls the substance of the communications that it receives.
>
> In other words, scientific research is a social activity. (Ziman, 1968, pp. 9–10)

Both science and democratic politics must rely on arguments capable of convincing their publics. The political public must continually decide issues for which information is lacking or contradictory. Majority rule is the most acceptable mechanism for deciding matters in these cases. The scientific public, on the other hand, decides an issue only when the evidence is strong and convincing enough to forge a universal consensus.

Ideally, a public is a collection of free, independent, and interested individuals, free to agree or disagree with the arguments, independent of social and economic pressures, and interested in the outcome of the debate. The scientific public consists of teachers, professors, and researchers in schools, universities, and laboratories around the world. The world being what it is, however, many scientists are strongly dependent on their employer or government for their livelihood, and so the scientific public isn't ideal. For example, the overwhelming evidence of the role of cigarettes in causing cancer may never convince a scientist working for the tobacco industry. Fortunately, most scientists don't work for the tobacco industry or any other single industry or government, and so their independence, even though not ideal, doesn't usually hamper science.

With this understanding, then, we can also define science as the study of those things about which the scientific public can form a consensus. This isn't a circular definition, since the scientific public exists before the attempt to establish a new fact or principle. To be taken seriously as science, any new idea or assertion must, at the very least, be capable in principle of convincing the established scientific public.

Put in this way it sounds as though there is no objective truth to science. Am I saying that science is just whatever the scientific establishment says it is? Most people, and certainly most practicing scientists, believe that there is more to it than that.

But think carefully about what has been said. A large public has been defined that is free, knowledgeable, and independent. It is not at all obvious that there should be any matters about which it would agree, let alone such esoteric ones as invisible atoms and helical genes. That there appear to be such things is our best evidence of an objective reality. Indeed, we can define objective reality as those matters about which the scientific public agrees. Ziman calls it public knowledge.

The earliest scientific public, that of ancient Greece in the sixth century B.C., consisted of only a few hundred "research-level" geometers and astronomers spread around the Mediterranean. They would seldom have met in groups of more than three or four, except perhaps in intellectual capitals such as Miletus and Athens. Communication was primarily by word of mouth, as students, traveling from one city to another, spread the ideas and theorems of their masters. Books would have played an important secondary role, because even though they were expensive to reproduce, they preserved knowledge from one generation to the next.

The members of a scientific public must share common knowledge, for example, the principles and methods of geometry, without being committed to a common ideology or school of thought. This generally requires geographical diversity, since the students of any one master tend to follow his ideological dispositions. Thus even though the number of scientists in ancient Greece was small, they formed a true public, spread, as they were, across the vast Hellenistic world.

The early revival of Greek learning in Europe in the thirteenth, fourteenth, and fifteenth centuries was too ideological to be termed scientific. But gradually, as the number of universities grew and their faculties gained some degree of intellectual freedom, a scientific public did emerge, and by the beginning of the seventeenth century it rivaled in size that of ancient Greece. Thereafter, spurred on by new discoveries and readily available books, it grew exponentially. By the end of the century, the number of scientists was in the thousands. Indeed, more scientists were born in the seventeenth century than had ever lived before.

Even more dramatic than the increase in the number of researchers was the increase in the number of possible contacts between researchers. With 100 scientists, there are about 5,000 possible two-person contacts; with 1,000 scientists, there are about 500,000 two-person contacts. A 10-fold increase in scientists results in a 100-fold increase in their possible contacts.[1] Thus the number of contacts among the scientists of the seventeenth century vastly exceeded the total number of scientific contacts that had occurred over all of previous human history.

These numbers have continued to grow. There are millions of scientists alive today, as many as have died since Pythagoras. There are 40,000 scientific journals published throughout the world, with at least 40,000 editors. The number of science editors alone exceeds the total free population of Athens—men, women, and children—at the time of Pericles.

The most obvious effect that this increase in numbers and contacts had on seventeenth-century science was a sharpening of the issues. The seventeenth century didn't invent experimentation, but it did refine the process. It became standard then to keep records of one's experiments, a depressingly difficult task that we must force our students to perform as they wrestle with the perplexities of a physics or chemistry laboratory course.

The larger scientific public of the seventeenth century, with its many sources of information, became ever more critical of loose and sloppy data. There was sure to be someone to spot a flaw, no matter how subtle, and announce it to the world. This process is the routine business of science today, usually performed out of sight of the public. After I once read a fatally flawed paper at a physics meeting, my fundamental error was whispered in my ear by a senior scientist, politely sparing me public embarrassment. Most of the time, one isn't so fortunate. The strange case of Martin Fleischmann and Stanley Pons, which is discussed in more detail later, gave the general public a rare view of this process in full force.

In Chapter 1, I proclaimed the certainty of much of our scientific knowledge. Such a bold statement is startling in itself, but even more so when one understands how strongly the erroneous views of Aristotle were adhered to for 2,000 years. How can I be so certain of, say, Newtonian mechanics, which has been around for only 300 years?

The answer is in the numbers. More scientists with better instruments examined Newtonian mechanics in the generation following the publication of the *Principia* in 1687 than had ever critically examined Aristotle in the preceding 2,000 years. Since then, this scrutiny has increased a millionfold. Newtonian mechanics is built into the fabric of our technology, from the computer codes that govern the flight of spaceships to the inertial guidance systems of submarines.

Again, let me emphasize, I am *not* saying that truth in science is whatever the scientific consensus says it is. Rather, I am saying that, in general, there can't be a consensus of informed opinion unless there is something "real" out there and that this consensus is our only objective proof of that reality. Because modern science is so new, historians such as Thomas Kuhn (1970) have concentrated on the early shifts and swings in that opinion as it groped for a deeper understanding of mechanics, heat, and light. And such shifts and swings will continue as science moves into unknown areas such as the workings of the human mind and the structure of the universe. But what millions of scientists today accept as established facts—atoms, gravity, electromagnetism—won't change in 1,000 years or 100,000 years.

Ziman's social definition of science helps explain how certainty can arise from a purely human activity. There is no cynicism in it. On the contrary, public (objective) knowledge is the greatest of all human inventions because it breaks the barrier of egocentrism that has imprisoned humankind throughout its existence. The test of public consensus is severe. Private insights and intuitions, no matter how strongly felt, won't do. Most of the traditional knowledge that has guided and misguided humankind from its beginning won't do.

Private Knowledge

Thomas Blackburn, a serious critic of science, believes that science as it is currently formulated is incomplete because it has no place for "direct sensuous experience, subjectivity, and respect for intuition" (1971, p. 1003). He believes in the ancient notion that the self is somehow so connected to nature that by being sensitive enough, it can gain direct knowledge of nature. This, of course, is just the egocentric identification of the self with the universe.

Science doesn't ban egocentrism, intuition, or subjectivity. It can't because they are intrinsic parts of human nature. Science doesn't care how a scientist comes up with an idea: by hard work, in a dream, or from a fortune-teller. It does care, however, about the evidence the scientist uses to support the idea. It must be convincing to those who don't believe in Ouija boards, not just to those who do.

But critics like Blackburn want more than this. They want intuitive "knowledge" based on private insight admitted as evidence. Were he to have his way, science journals would have to open their pages to all the mystics and visionaries of the world. Science would be lost in the babble of conflicting claims of "truth."

Scientists and laypersons alike are incensed when they realize that science vigorously censors what appears in a scientific journal. Every article must be approved by several peer reviewers. These reviewers are researchers themselves, whose own research passes through the same process. The main purpose of the review is to check whether the paper is up to the current standards of the field. Is the author aware of the latest ideas and methods? Are there any obvious mistakes in logic or methodology? Are other related works properly referenced? Does the paper really have anything new to say? In short, is the paper capable of affecting the consensus in the field?

A revolutionary insight disconnected from the current consensus is unlikely to be published. For example, the first paper on strong focusing, a technique used today in all high-energy particle accelerators, was rejected several times. The author, a Greek engineer, had written it in terms that were unfamiliar to the American physicists who reviewed it. It took a year before someone recognized the importance of the work, and the necessary changes were made to make the paper understandable to its intended audience.

Velikovsky

The situation was quite different in the case of Immanuel Velikovsky. His best-seller *Worlds in Collision* hypothesized that near collisions of Venus and Mars with the earth had caused many of the miracles mentioned in the Bible (Velikovsky, 1950). His ideas, which contradict the well-established principles of Newtonian mechanics, were rejected outright by the scientific community. Macmillan, which originally published the book, transferred its rights to Doubleday under pressure from several prominent astronomers, an action that led to the charge of censorship being leveled against the scientific community (de Grazia, Juergens, and Stecchini, 1966; Goldsmith, 1977).

There seems to be a great deal of confusion on this point. Publishers, or their divisions, tend to specialize in certain fields. Within a field, they rely on their reviewers to advise them on the appropriateness of prospective books. In the case of *Worlds in Collision,* Harlow Shapley, a prominent astronomer, very properly advised Macmillan of the possible adverse effect its publication of an extremely unorthodox work would have on its reputation. Nevertheless, Macmillan did publish it. Only after the predicted fire storm erupted over the book did it heed Shapley's advice and transfer the book to a more appropriate publisher.

This wasn't censorship in any useful meaning of the word; it was an attempt to prevent a blatantly unscientific work from being published with the scientific imprimatur that the Macmillan name carried. Velikovsky received high praise in the literary magazines, being compared with Einstein, Newton, Darwin, and Freud. He became a best-selling author and lecturer; his many fanciful theories were published in a succession of books; and he was in demand on campuses around the country.

What he couldn't do was get an article accepted by a scientific journal. Scientific journals must remain the preserve of articles capable of affecting the consensus of the scientific public. Books are the place for opinions, speculations, and fanciful accounts of ricocheting planets. The publisher has only to convince enough buyers to cover their cost of publication. In a free society with a vigorous press, there is little danger that an important idea will not get a fair hearing.

To have an article accepted by a scientific journal, Velikovsky would have had to have given a noncontradictory account of his theories. He claimed, for instance, that when Joshua caused the sun to stand still, the earth did in fact stop rotating. If the earth had slowed down gradually, say over an hour or two, the change in motion might not have been noticeable. It couldn't have taken much longer, because then the longer day would have been noticed worldwide. But the braking of the earth's motion in a few hours would have generated enough heat to melt the earth's crust—if the heat were concentrated on the earth's surface—or to raise the total temperature of the earth to the boiling point of water—if the heat were distributed uniformly throughout the earth's interior. In either event, the catastrophe would have destroyed all living things.

Velikovsky, of course, had an answer for this. Maybe the earth didn't stop rotating. Maybe its axis tilted in such a way that a retrograde motion was produced that had the effect of making the sun appear to stand still. OK, how much was the tilt? Would it have had the visual effect claimed? Velikovsky never said.

It is virtually impossible to refute such ill-formed ideas by scientific means. As Carl Sagan says at the conclusion of his analysis of *Worlds in Collision:*

> To the extent that scientists have not given Velikovsky the reasoned response his work calls for, we have ourselves been responsible for the propagation of Velikovskian confusion. But scientists cannot attempt to deal with all areas of borderline science, of which the number is legion in America today. The thinking, calculation, and preparation of this chapter, for example, took badly needed time away from my own research. But it was certainly not boring, and at the very least I had a brush with many an enjoyable legend. I hope that in the future Velikovsky's views and the views of other popular proponents of borderline science will receive a reasoned, if hopefully briefer, scientific response. (1977, p. 93) [2]

I might add that there are few scientists with the knowledge, patience, and literary skill of a Carl Sagan. And even Sagan uses more mathematics and technical terms than would be acceptable in, say, *Harper's Magazine,* which published a condensation of *Worlds in Collision* in January 1950. Well-written pseudoscience, with its exciting generalizations and lack of mathematics, can always find a bigger audience than can carefully crafted, but necessarily tedious, rebuttals.

Science is the search for consensus. But before there is sufficient evidence to form a consensus on a particular problem, scientists will pursue different approaches and suggest different solutions. Passions may run high as the conflicting ideas of strong-willed individuals struggle for dominance. But science, like parliamentary democracy, depends for its survival on individuals who can subordinate their egocentrism to a higher ideal. Democracy isn't possible in societies in which the losers resort to violence and guerrilla warfare, and science isn't possible when researchers, failing to gain acceptance for their ideas, break away from their colleagues to form rival schools.

Parapsychology

Chiropractic, homeopathy, psychoanalysis, and parapsychology are examples of rival schools that formed when their founders couldn't convince competent colleagues of their ideas. The first three are therapies that are sold directly to the public without any reasonable theoretical basis or any objective evidence of their efficacy. Parapsychology has always had a tenuous claim to academic respectability, as its methods mimic those of conventional psychology. It can be distinguished from science, however, by its refusal to acknowledge the need to achieve consensus.

As a boy at camp, my bunkmates and I amused ourselves at times with experiments in mental telepathy. One of us would stare at a card while another would try to determine it by reading the starer's mind. With an ordinary deck there is only one in fifty-two chances of a correct read, but we did get one early in our "research," which greatly encouraged us. We were also pleased when we got a near hit, perhaps a card of the same value but a different suit, or a card with the same color and nearly the same value, and so on. I gradually lost interest when I realized that our definition of "near" was so loose and our record keeping was so sloppy that we were proving nothing. Nevertheless, some of my friends persisted, haunted by our early success.

I thought little about the matter for several years, but when

J. B. Rhine, a famous parapsychologist, spoke at the University of Wisconsin during my freshman year, I was an eager listener. He started with a discussion of his experiments on mental telepathy, which gave a very nice positive effect. Naturally, I was inclined to believe the results of this Duke University professor who had been invited to the Midwest to present them.

But then Rhine went on to discuss clairvoyance, in which the subject guesses the cards as they are taken from the top of a shuffled deck. No one looks at the card until after the guesses are completed. Since there is no physical way for the information about the card to reach the subject, the experiment must give a null result. It would be an ideal control to test Rhine's methodology on the telepathy tests. To my surprise, Rhine claimed positive results for these experiments, too. He claimed positive results when the cards were in the next room or in a room hundreds of miles away. He claimed positive results even when the subject guessed one day and the cards were shuffled and selected the next.

Even as a freshman I could see that Rhine's positive results for these clairvoyant and precognition experiments invalidated his telepathy results. If nothing makes the effect go away, then the effect must be attributed to a flaw in the method. An experiment from freshman physics will illustrate my point.

Picture a small metal tube connected by an electrical wire to a metal box with dials, meters, and a speaker that is emitting strange clicking sounds. On the table next to the tube is a plastic disk about one inch in diameter. I claim that inside the plastic is a small quantity of radioactive material and that the clicking sounds are caused by invisible rays emanating from this material and entering the tube. How do I persuade someone who knows nothing about radioactivity or Geiger counters that there is a connection between the disk and the clicking sounds?

The answer is obvious. If I (or, better, you) move the disk closer to the tube, the rate of clicking will increase; if you move the disk away from the tube, the rate will decrease. If you place a sheet of lead between the disk and the tube, the rate will decrease; when you remove the lead, the rate will return to normal. These simple experiments demonstrate very persuasively that there is a cause-and-effect relation between the disk and the tube and that something physical is traveling from the disk to the tube that is partially blocked by lead.

To see just how persuasive these experiments are, suppose that the rate didn't change when you moved the disk closer to or farther from the tube or when you placed lead between the disk and the tube. What would you conclude? That the emanation, whatever it is, doesn't weaken with distance or passage through lead? Or that the clicking sound was due entirely to the electronics inside the box? Certainly, if you had no reason to believe that the disk affected the tube, your opinion wouldn't be changed by the fact that the clicking rate remained the same in all circumstances.

But suppose, for some egocentric reason, I was convinced that the disk affected the tube, even when moving or blocking the disk didn't change the clicking rate. I could then claim the discovery of a new emanation that didn't weaken with distance or passage through lead. Most sensible people wouldn't believe me, of course, but being a charismatic guy, I could probably convince a few students of my views. We would form a new field, start our own journal, and berate the establishment for being closed minded. But we wouldn't be doing science because we wouldn't be using methods capable of convincing all competent observers.

Joseph B. Rhine began his studies in extrasensory perception (ESP) at Duke University in 1930. Most of these were done with cards with one of the five ESP symbols: circle, cross, wavy lines, rectangle, and star. Usually a deck was used that had five cards of each kind, twenty-five in all. The deck was shuffled and cut, and the subject called the cards in the order they were picked from the top of the deck. Since there are five different cards, there is a one-in-five chance of a correct call, or "hit." In twenty-five calls, one should get, on the average, five hits ($1/5 \times 25$), but variations of plus or minus four aren't unlikely. That is, it wouldn't be the least surprising if you got nine hits (out of twenty-five) one time and two another.

Thus the effect can't be established casually. Extensive testing is required. It isn't remarkable to get 6 hits (out of 25) in 25 calls or even in 100 calls. Thousands of calls are required under carefully monitored conditions. The experimental conditions of Rhine's early work at Duke were rather loose, to say the least. Here is his description of a subject who consistently scored 6 (out of 25) in 7,500 calls:

> [Charles E.] Stuart was a subject of Dr. Lundholm and me in the trance-telepathy series. He scored, at that time, just a little above chance average. That is the level also of his first 100 trials witnessed by me alone, or more exactly at an average of 6 hits per 25 calls (with 5 expected). The other 40 trials I have

witnessed raised the average for the whole 140 trials to 6.15
hits per 25. All the rest of his work is unwitnessed but, since
he does not on the whole rise beyond the level of these witnessed results and since he is the responsible man he is, I feel
that we may unhesitatingly offer his work to the public as fully
worthy of consideration.

Beginning in the autumn of 1931, and continuing through
the school year, Stuart ran the huge sum of 7,500 trials. Marvelous patience indeed! [His positive results] were not very
high but, on account of the large number of trials, they take on
great mathematical significance. The ratio of the positive devia-
tions to the [probable error] rises to 13.5. The average per 25
for 7,500 was 6.05. . . . In his procedure Stuart held the cards
behind him, cut the pack there at the start, and held each card
by the corner between thumb and finger, recording each call
when made and checking up after every 5 calls (and then re-
shuffling). (1934/1964, p. 92)

Has Rhine made a reasonable case? Certainly one must be bothered
by the fact that Stuart did the tests on himself without witnesses.
But Rhine faces the problem directly, arguing that Stuart's private
data are reliable because they show the same rate of success as do
the witnessed data. Allowing for the possibility of self-deception and
even deceit on Stuart's part, the average reader would probably de-
clare the results suggestive but not definitive.

Such open-mindedness is an appropriate way to handle much of
the uncertain and conflicting information that confronts us. However, because Rhine claimed his work was science, the science community couldn't be open-minded about it. In the first place, journal
editors were faced with the immediate decision of whether Rhine's
work was bona fide science. In the second place, scientists at large
couldn't be open-minded about the existence of such manifestly
noncausal phenomenon as clairvoyance and telekinesis.

Probably nothing disturbs nonscientists about science more than
its seemingly dogmatic rejection of claims of the paranormal. But
science isn't rejecting the claims themselves so much as the evidence
used to support them. Scientific evidence, by our definition, must be
strong enough to win a consensus. That is an exacting standard. The
scientist, like a stage magician, can't cover his hands at a critical part
of the demonstration. The audience would boo and throw tomatoes.

The procedural irregularity of Stuart's unwitnessed tests on
himself is enough to discredit the work. Rhine's ingenuous sugges-

tion that we trust Stuart because "he does not on the whole rise beyond the level of these witnessed results and since he is the responsible man he is" exemplifies the faulty reasoning behind all of Rhine's work. The point is that Stuart's positive result on his 140 witnessed trials isn't statistically significant, whereas the same result (6 out of 25) on 7,500 trials is. That is, his witnessed result is within our expectations of no ESP, whereas his unwitnessed result isn't. The witnessed result can't, therefore, be used to validate the unwitnessed result.

Rhine's sloppy methodology was severely criticized after his first book, *Extra-Sensory Perception,* was published in 1934. In response, he reluctantly tightened up his procedures and never again obtained the spectacular successes of his early years. But neither did he retract this early work. For the 1964 reprint of *Extra-Sensory Perception,* he wrote:

> But can we take these early tests of the 1930s seriously? If there has been this long period of debate over the adequacy of test procedures, may not these early experiments have been so loosely conducted from today's point of view as to be relatively worthless? No, and I say it with emphasis! . . . As one looks back, he wonders continually, "Why did we not see such and such a weakness?" . . . No one will ever know.
>
> . . . [T]his early report . . . tells . . . how long we were able to keep those early subjects scoring well in spite of the monotony of the procedure, how well the subjects responded to the new conditions introduced, and what sort of program we had that kept so many so productive for so long. What would we not give today for the like of that?
>
> . . . [W]e . . . read this book today . . . to know what made [the evidence for ESP] so good then—so good in terms of performance level, so good at this crucial beginning. (1964, pp. xxxv–xxxvi)

This passage reveals the fundamental difference between science and pseudoscience. In science, all knowledge is public knowledge, that is, knowledge based on evidence strong enough to convince essentially the entire scientific public. In pseudoscience, private knowledge based on personal insight, intuition, or belief is confused with public knowledge.

This confusion is seen in Rhine's acknowledgment, on the one hand, that his early experiments were "loosely conducted," and his claim, on the other, that these experiments provided strong evidence for ESP. An experiment is deemed loose when it is so poorly controlled that its results can't be convincingly attributed to any one specific cause. Stuart's results, for example, could have arisen from any number of causes, such as fraud, data selection, and shuffling irregularities,[3] each one of which is considerably more likely than ESP. Rhine's rejection of the more likely causes for the least likely is a matter of his private judgment. This, by itself, is merely an idiosyncrasy. It becomes pseudoscience when by claiming that the experiment is strong evidence for ESP, he offers his private knowledge for serious public consideration.

But why must the results of a loose experiment be rejected out of hand? Isn't it possible that the results are still valid, even though the experimenter didn't dot every *i* and cross every *t?* This last question seems profound only because it's ambiguous. If the question is "Might not ESP be true, even though the experiment didn't prove it?" the answer is obvious: "Maybe, but this experiment, having proved nothing, deserves to be rejected." If the question is "Couldn't ESP be the real cause of the results of the experiment, even though other causes weren't ruled out?" the answer is the same: "Maybe, but having failed to rule out other reasonable causes, the experiment proves nothing, and should be rejected."

Such a hard-nosed view of evidence isn't to everyone's liking. Science's exclusion of personal knowledge offends the cherished egocentric belief that one can have a direct personal relationship with nature or the supernatural. Blackburn frankly criticizes science for this rejection. For Rhine, psychic phenomena are the result of just such a relationship. The need for personal knowledge runs deep in the human psyche.

The knowledge that science offers instead isn't impersonal but public. It's shared knowledge, the common heritage of all humankind. Human beings may pray to different gods, sing different songs, and practice different rituals, but they all generate their electricity in the same way.

Science must be skeptical, not open-minded. There are millions of claims in the world, and it isn't possible to prove any one of them false by scientific means. I can't prove, for example, that there are no invisible beings from Mars in my basement, or closet, or word processor. No matter how many tests I perform that fail to verify their existence, I can't prove to a believer that some test I didn't try may detect them. Therefore, the burden of proof must be on the believer. If the evidence is convincing enough, the skeptics will in time accept almost anything, even that the continents are drifting about the face of the earth. But until the evidence is there, the only sane course is to reject all claims that are unverified and inconsistent with current knowledge.

After more than ten years of negative results in parapsychology, Susan Blackmore had to face the question of open-mindedness. She got hooked on the occult at Oxford University, where, while majoring in physiology and psychology, she spent her spare time playing with Ouija boards, exploring graveyards, and experimenting with ESP. A true believer but a trained scientist, she spent ten years performing experiments to demonstrate psi, the elusive "principle" of ESP. All her results were negative:

> "What do these negative results tell us?" Of course the one thing they do not tell us is that psi does not exist. However long I went on looking for psi and not finding it they could not tell us that. But I found myself simply not believing in psi anymore. I really had become a disbeliever. Like one of those doors with a heavy spring that keeps it closed, my mind seemed to have changed from closed belief to closed disbelief.
> . . . In the end I think my negative results told me that the psi hypothesis . . . forces us to ask ever more boring questions, culminating in the question "Does psi exist?" and to that question there is no obviously right answer. Where there is no right answer, we are in ignorance; and, where we are in ignorance, we should do only one thing—have an open mind. But that is too difficult. After all these years of research, I can only conclude that I don't know which is more elusive—psi or an open mind. (Blackmore, 1987, p. 249)

Blackmore is correct that negative results can't disprove anything and that one must be open-minded in the face of ignorance. She also poignantly describes how psychologically distressing true open-mindedness can be. But she is wrong in ascribing ignorance to our inability to disprove preposterous claims by means of experimentation. This inability is inherent in the logic of the empirical method. We can, however, point out the lack of evidence of such claims and demonstrate their logical inconsistency with everything we do know about the world. And about logically inconsistent things for which there is no evidence, we need not be open-minded. After all, there are only so many impossible things one can believe before breakfast.

Parity Violation

Science progresses incrementally. Each step must be carefully taken to preserve consistency with everything before it. This plodding approach may be at variance with the image of swashbucklers of the mind overthrowing established ideas with sudden flashes of genius. But because of the consistency requirement and the immense base of our current knowledge, such revolutions are rare.

Nevertheless, as recently as 1961 a major principle of physics was found to be wrong or, rather, not as universally applicable as had been thought. This was the conservation of parity, or indifference to reflection. Up to that time, all the established laws of physics, such as those for gravity, electromagnetism, and nuclear interactions, were found to be independent of absolute position, orientation, and reflection. The equations that describe these laws don't have any terms in them that contain a specific position (e.g., the center of the earth, the center of the sun, or the center of the galaxy), a specific direction (e.g., toward the center of the galaxy), or a specific handedness (e.g., left- or right-handed). This means that the laws of physics are the same in all parts of the real world as well as in the reflected world behind the looking glass. (Indifference to reflection and handedness are equivalent, since the reflection of a right hand is a left hand; try shaking hands with your mirror image.) It was a sacred rule that any proposed new law of physics must be similarly indifferent to position, orientation, and handedness.

Such a rule is of great help in formulating new physical theories, because it restricts the possible forms that such theories can have. Any theory that depended on position, orientation, or handedness was simply not acceptable. This may seem unnecessarily closed-minded and dictatorial. But the fact is that all known theories met this restriction, and it would be chaotic to admit, even for consideration, new theories that didn't.

Yet in 1956, when the theorists T. D. Lee and C. N. Yang published a paper suggesting that indifference to reflection may not be a universal principle, they were taken very seriously. Experiments undertaken to check on their suggestion showed that indeed there were fundamental processes that were different in the real world than in the reflected world (Wu et al., 1957). Indifference to handedness, or parity conservation as physicists call it, was dead. Long live parity violation (Lee and Yang, 1956; Morrison, 1957).

This rapid switch in the established canons of physics was possible because Lee and Yang showed that no inconsistency would result. They carefully reviewed all the experimental evidence for parity conservation and found that although it was firm for processes involving electromagnetic and strong nuclear interactions, there was essentially none for processes, such as beta decay, that involve the weak interaction. They had undertaken this survey because discoveries in particle physics had indicated the existence of two elementary particles of the same mass that decayed in different ways. There were either two different particles with the same mass—a unique situation—or a single particle that decayed through a mechanism that violated parity conservation.[4]

Lee and Yang showed that if the latter was accepted, it wouldn't be inconsistent with previous experience, because there was, in fact, no direct evidence of parity conservation in such decay processes. They suggested an experimental test that involved placing radioactive cobalt in a strong magnetic field and cooling it to a temperature of one-hundredth of a degree above absolute zero. Difficult as this is, the experiment was immediately performed through the collaborative efforts of C.-S. Wu of Columbia University and E. Ambler and his associates at the U.S. National Bureau of Standards. The results dramatically established parity violation in weak interactions and earned Lee and Yang the Nobel Prize in physics the following year.

Consistency is the basic principle of rational discourse. Without consistency, there can be no consensus and hence no science. Turn the argument around, and we can say that the existence of science implies a consistent universe. Thus the view of science as public knowledge doesn't deny that science is based on the existence of an external world. But the sociological approach avoids having to postulate such a metaphysical concept at the beginning of the discussion.

It also helps explain why science is never as empirical as we would like to think it is. The story of the entire scientific community's giving up a basic principle on the basis of hard evidence would be more glorious were it not for the fact that parity violation had been observed thirty years earlier by Cox, McIlwraith, and Kurrelmeyer (1928) and Chase (1929). These investigators discounted their own observations of anomalous asymmetries because they violated an accepted principle. Only with the accumulation of additional evidence and the demonstration that this evidence wasn't in irreconcilable conflict with parity conservation in nuclear and electromagnetic interactions could parity violation in weak interactions be accepted.

Cold Fusion

The controversy in 1989 over the alleged discovery of cold fusion is a textbook case of the value of stubborn closed-mindedness in the face of experimental evidence. On March 23, 1989, B. Stanley Pons, a chemist at the University of Utah, announced that he and Martin Fleischmann, a chemist from the University of Southampton, England, had obtained a controlled fusion reaction in a small electrolytic cell. A group of physicists at Brigham Young University, headed by Steven Jones, immediately announced similar results. The two groups had been working independently of each other but had become aware of each other's work some months earlier. At a meeting that involved the presidents of these two Utah universities, an agreement was reached that on March 24, 1989, each group would submit a paper on its work to *Nature,* the prestigious journal of the British Association for the Advancement of Science. Pons and Fleischmann also announced their results at a press conference on the same day, which wasn't part of the agreement.

The announcement of scientific results at a press conference, prior to their publication, is frowned upon. A paper isn't even deemed publishable until it has undergone critical review by other researchers in the field. This occurs after the paper is submitted to a journal and often results in the paper's being returned for revision. Methodological errors are common; although they are often overlooked in experiments that are consistent with expectations, they can be lethal for an experiment that reputes to have discovered something inconsistent with previous experience. *Nature* did return the Fleischmann and Pons paper but accepted the Jones group's paper (Jones et al., 1989). Fleischmann and Pons had already had their cold-fusion paper accepted by the *Journal of Electroanalytical Chemistry* (Fleischmann, Pons, and Hawkins, 1989).[5]

How inconsistent were the Utah results with previous experience? Fusion reactions—nuclear reactions in which two light nuclei combine to form a heavier one—have been intensely studied for over fifty years. They are the basis of the hydrogen bomb, of the energy production of the sun, and of research efforts to produce controlled fusion on earth. The Utah experiments tried to fuse two deuterium (hydrogen-2) nuclei. When two deuterium nuclei fuse, they form a tritium (hydrogen-3) nucleus and a proton, a helium-3 nucleus and a neutron, or a helium-4 nucleus and a gamma ray. The gamma-ray reaction is rare, occurring only once for every hundred occurrences of the neutron and proton reactions.

A deuterium nucleus consists of one proton and one neutron; it is an isotope of the hydrogen nucleus, which is just a single proton (see Chapter 1). Normally the deuterium nucleus has an electron encircling it, in which case it is an atom of deuterium (heavy hydrogen). Deuterium, like hydrogen, is a gas at room temperature; in this gas the deuterium atoms are combined in pairs, forming a molecule.

The difficulty of achieving fusion comes from the fact that two deuterium nuclei repel each other electrically because each is positively charged. They normally don't get close enough to interact. In a deuterium molecule, the electrons overcome this repulsion to a large extent, and the two nuclei in the molecule are separated by only three-billionths of an inch (seven-billionths of a centimeter). They are still too far apart to fuse, but Jones and his colleagues had previously achieved fusion by replacing one of the electrons in a deuterium molecule with a negatively charged muon (Rafelski and Jones, 1987). A muon is 200 times heavier than an electron and so brings the two deuterons 200 times closer together. At this separation, fusion occurs at a measurable rate, though not enough energy is released to pay for the cost of creating the muons.

It thus is clear that cold fusion is possible if two deuterium atoms can be squeezed together closer than they are in a deuterium molecule. It is also well known that many metals, including palladium, absorb hydrogen. Therefore, isn't it reasonable to suppose that if deuterium were forcibly incorporated into palladium by using an electrical current, deuterium atoms could be squeezed together close enough for their nuclei to fuse?

No, it isn't. The palladium atoms are themselves three times farther apart than are the two deuterium atoms in a deuterium molecule. The palladium is able to absorb deuterium molecules *because* the spacing between the palladium atoms is larger than the diameter of the deuterium molecule. No squeezing is involved. In fact, the deuterium molecule breaks apart inside the palladium, and its two deuterium atoms end up being farther apart in the palladium than they were in the free deuterium molecule.

Furthermore, Fleischmann and Pons claimed that their fusion

reaction generated a large quantity of heat. A simple calculation shows that if the heat they claimed was due to fusion, there would have been enough neutrons generated to have killed the experimenters. They interpreted the absence of the neutrons as the discovery of a new type of nuclear reaction.

Scientists weren't immediately aware of all this when the announcement was made at a press conference. So when reporters asked scientists for their assessment of the Utah experiments, there were mixed responses. Philip Morrison observed, "Based on the information I have, I feel it's a very good case." He said his confidence in the reality of the reaction was "high, but not conclusive." The Nobel Prize–winning physicist Sheldon Glashow said, "I don't believe a word of it" (Chandler, 1989a, p. 5). The amusing comment by Kim Molvig—"I am willing to be open-minded, but it's really inconceivable that there is anything there" (Pool, 1989, p. 1661)— probably reflects the ambiguous use we often make of "open-minded." Most alarming were the comments from scientists that put extraordinary confidence in Fleischmann and Pons: "I'd be extremely surprised if they've done anything stupid. They have a very good track record in electrochemistry. I am pretty excited about this" (Chandler, 1989b, p. 29). In fact, stupidity in science is far less surprising than is the radical overthrow of well-established doctrine.

I was one who was convinced from the beginning that the results were bogus. You could say I was closed-minded on the subject in the sense that I had a deep inner conviction that the experiments were wrong. Within days of the announcement, we had an impromptu meeting in my department to discuss the matter. By then, prepublication copies of the papers of Fleischmann and Pons and the Jones group had arrived over the fax machine. As I examined them, I judged the first unsuitable for publication and the second an interesting and publishable work that was probably wrong. My judgments at that time were more negative than those of my colleagues but were in accord with those of the reviewers at *Nature,* who rejected the Fleischmann and Pons paper while accepting the Jones group's paper.

In their original press conference, Fleischmann and Pons stated that they had been working on cold fusion "in secret" for five years. This tipped me off at once. Science isn't done in secret. It's too easy to make a mistake. This is true for electrochemists doing electrochemistry and even truer for electrochemists doing nuclear physics. A theoretical nuclear physicist myself, I have a small idea how hard my experimental colleagues work to get everything right. It's a process of constant refinement in which comments and criticisms of other scientists play a vital role. To me, Fleischmann and Pons were acting like a couple of alchemists for whom the scientific revolution had never happened. The public, on the other hand, and many scientists as well, were taken by the romantic notion of two chemists working alone to defeat the billion-dollar hot-fusion physicists. The chemists ate it up.

As soon as I saw the Fleischmann and Pons paper, I knew my worst suspicions were correct. The paper lacked any references to nuclear physics and, indeed, had hardly any references at all. References are used in all scholarly writing to establish the author's knowledge of the relevant work in the field. A typical scientific paper has about thirty citations, most to recent related work. Fleischmann and Pons had five references, three to their own work and two to works on hydrogen absorption in metals. (One of these was to a 1940 Russian paper.) There was no reference to the huge literature on fusion. A scientific paper differs from a nonscientific one by the requirement that it be consistent with the previous work in the field. This means it must acknowledge the existence of this previous work and show how the new results fit in with the old. A paper may challenge previous work, but it can't ignore it.

The Jones group's paper has eighteen references. They indicate their knowledge of the unlikelihood of cold fusion, but argue that something unusual happens when the hydrogen is absorbed into matter. They interpret some geophysical data as indicating that cold fusion is going on inside the earth. This connects their work to the rest of science. They also give their raw data—the number of neutrons they detect when their cell is off and when it is on. The difference, if any, can be attributed to fusion. Their data are shown in Figure 8.1. The vertical bars on the points indicate the probable error; that is, the true value is likely to lie somewhere on the bar. To accept cold fusion, you have to believe that the positive results around the 100-channel mark isn't a statistical fluctuation. This I find very difficult, given the Jones group's claim that the neutron production "dropped dramatically" after eight hours. Statistical fluctuations are well known to produce dramatic changes in small numbers; that's what makes gambling so exciting. Many other laboratories have since tried to duplicate the Jones group's result. Although some claimed to see a few neutrons, none of these results has held up over time.[6]

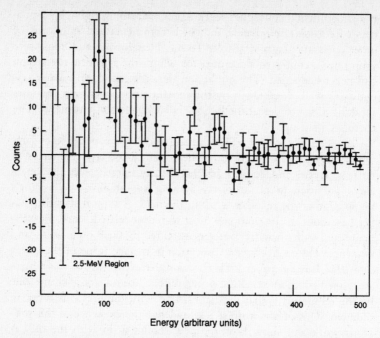

Figure 8.1 The number of neutrons of different energies detected by Jones et al. with the electrolytic cell on, minus the number detected with the cell off. There is a positive difference in the energy region expected for fusion neutrons (2.5 MeV). Although this difference is statistically significant on its face, such results are sometimes found to result from a bias in the way the data are handled. (Reprinted with permission from S. Jones et al., "Observations of Cold Fusion in Condensed Matter," *Nature* 338 [1989]: 737)

The fact that two Utah research groups both had reported cold fusion at the same time greatly confused the picture, as they seemed in some way to confirm each other. However, the most casual reading of their papers, or of the newspapers at the time, clearly showed that the two results were highly inconsistent with each other. The small fusion rate indicated by the neutrons detected by the Jones group, though inconsistent with current nuclear and solid-state theory, isn't as inconsistent as the trillion times higher rate reported by Fleischmann and Pons without the production of the necessary neutrons.

The Fleischmann and Pons paper, for all its extravagant claims, offers no data. Although Fleischmann and Pons measured the temperature change in their fusion cell, they give none of their raw readings. They report only highly processed numbers, without giving any indication how they were calculated. When they saw only one-ten-trillionth (10^{-13}) of the neutrons that their claimed fusion rate should have produced, they claimed the discovery of "an hitherto unknown nuclear process or processes." When they came in one morning and found their cell burned out, they claimed that fusion did it.

I was surprised how many scientists were prepared to relax their usual skepticism in the face of such an unscientific paper. The chemists were particularly excited, and at a meeting they had great fun claiming that two of theirs had done in five years and with a few thousand dollars what hundreds of physicists with billions of dollars haven't been able to do in forty years. A little team spirit is a good thing, but this was ridiculous. Back home, I was lecturing my physics students on the absurdity of the Fleischmann and Pons claims while their chemistry professor was excitedly expounding on the importance of the Utah result.

There are many lessons from this episode. First, scientists themselves are often poor judges of the scientific process. This isn't as surprising as it seems, since their training is purely technical. Many don't appreciate the seriousness of violations of procedures, such as giving a press conference before publication. More surprising, many took Fleischmann and Pons's incredible conclusions about their own work at face value, even before reading their paper.

Second, scientific research is very difficult. Anything that can go wrong will go wrong. For instance, Fleischmann and Pons forgot to stir their cell while measuring its temperature, thereby totally invalidating their measurements. Stirring is a routine procedure, taught to freshman in their first heat experiment, but neglected by a Fellow of the Royal Society working with the chairman of the Chemistry Department of the University of Utah. This isn't to criticize their professionalism; it would happen to any small group working in isolation on a fixed idea. No one, no matter how experienced, can do a complex experiment without the guidance and criticism of others. Isolation is the death of science.

Third, science isn't dependent on the honesty or wisdom of scientists. As a collective enterprise, it rises above individual shortcomings. It will survive Fleischmann and Pons, but only after the wasteful expenditure of hundreds of man-years of work and at least one death.[7]

Pathological Science

There have been other cases like cold fusion, but not as widely reported. The Nobel Prize–winning chemist Irving Langmuir (1881–1957) described some of them in cautionary talks he gave on pathological science. His talk on December 18, 1953, was recorded, and the transcription made of it by Robert Hall circulated privately for years before being published in 1989 (Langmuir, 1989). I shall retell two of Langmuir's tales of pathological science and add one of my own.

The discovery of X-rays by Wilhelm Roentgen in 1895 was a major event in science. While most scientists were content to learn as much as possible about this mysterious new emanation, others wanted the glory of discovering an emanation of their own. So perhaps it wasn't surprising that in 1903, Prosper René Blondlot, a distinguished member of the French Academy of Sciences, announced the discovery of N-rays, which he had produced by heating a wire inside an iron tube. Although these rays didn't pass through the iron, they did pass through an aluminum window in the iron. They were detected by looking at a very faintly illuminated screen in an otherwise dark room. If N-rays were present, the screen became more visible; of course, a great deal of skill was needed for this, because the screen was just on the edge of visibility. Under these conditions, Blondlot discovered that many different things give off N-rays, including people. Then he discovered negative N-rays, which decreased the visibility of the screen. He published many papers on the subject, and so did many others, confirming a multitude of unusual properties for these rays.

Among them was the discovery that N-rays could be broken up into a spectrum by passing them through a large aluminum prism. In 1904, the American physicist R. W. Wood visited Blondlot and found him measuring, to one-tenth of a millimeter, the position of the N-rays as they came through an aluminum prism. "How is that possible," Wood asked, "when the original beam is coming from a slit two millimeters wide?"

"That's one of the fascinating things about the N-rays," Blondlot replied. "They don't follow the ordinary laws of science." Wood then picked up the aluminum prism that was bending the N-rays onto Blondlot's screen and put it in his pocket. Blondlot was unaware of this because the room was very dark, and he continued measuring the spectrum to one-tenth of a millimeter. Wood published a report of this incident, which put an end to N-rays.

The point, of course, is that the observations were made at the edge of visibility. Under the conditions of the experiment, one could see or not see whatever one wanted. The observer wasn't separated from the thing observed, and so objectivity was nonexistent.

Langmuir found that the cases of pathological science that he had studied shared certain characteristics. They are generally claims, based on a weak or marginal effect, of fantastic phenomena contrary to all experience. There are conflicting reports from independent investigators. Reasonable explanations of the data, based on known science, are rejected. Interest rises rapidly for a time and then gradually fades away.

I know of one business consultant who won the extra confidence of his clients when, at the height of excitement about cold fusion, he correctly told them that there was nothing to it. This was, in fact, an easy call. Real discoveries of phenomena contrary to experience are very rare, whereas fraud, fakery, and foolishness are all too common. Thus, a closed-minded "I don't believe a word of it" is going to be correct far more often than not.

But in addition, cold fusion had the classic signs of pathological science. Its claims were certainly contrary to all experience or understanding of both nuclear and solid-state physics. Even the small number of fusion neutrons seen by the Jones group is fantastic. The much larger amount of fusion claimed by Fleischmann and Pons, without an accompanying lethal dose of neutrons, is doubly fantastic. A miracle squared.

A number of confirming experiments were reported soon after the cold fusion was announced, followed by a deluge of nonconfirming experiments. At one point, we had the peculiar situation that physicists largely rejected, while many chemists accepted, the cold fusion claims. In July 1989, a U.S. Department of Energy panel stated that there wasn't sufficient evidence of cold fusion to warrant government funding. Still, Fleischmann and Pons, together with the remnants of their believers, carried on, supported for a while by the largess of the state of Utah.

The neutrons measured by the Jones group are a sure sign of a nuclear process. However, they saw very few of them, and these mostly in the first hour of the experiment. There has been some confirmation of this, but many groups haven't seen any more neutrons than are usually present in background radiation. So this is clearly a marginal effect. The heat claimed by Pons and Fleischmann—though not in principle marginal—in fact resulted from heavily processing data obtained by incorrect procedures. They never actually generated the heat they talked about.

But for the layperson, unfamiliar with the science involved, the most characteristic sign of pathology is the language used. Fleischmann and Pons's insistence that "it is inconceivable that this [heat] could be due to anything but nuclear processes" isn't the language of science. It is the language of minds fixed on their own egocentric beliefs and unwilling to respect the justified skepticism of others

In 1934, Langmuir himself visited the parapsychologist J. B. Rhine at Duke University and pointed out that Rhine's work had all the characteristic symptoms of pathological science. Rhine thought it would be great if Langmuir published this. "I'd have more graduates," he told Langmuir. "We ought to have more graduate students. This thing is so important that we should have more people realize its importance. This should be one of the biggest departments in the university."

Langmuir spent the whole day with Rhine, who was in a philosophical mood. "People don't like these experiments," he said. "They don't like me. Sometimes, to spite me, they made their scores purposely low. . . . [These low results] I took and sealed in envelopes and I put a code number on the outside, and I didn't trust anybody to know that code. Nobody."

Langmuir thought this interesting: "You said that you had published a summary of all your data and that the average was 7. Now you are saying you have additional data that, if added to your published data, would bring the average to 5. Will you do this?"

"Of course not," he said. "That would be dishonest. . . . The low scores are just as significant as the high ones, aren't they. They proved that there is something there just as much, and therefore it wouldn't be fair [to combine negative and positive data]" (Langmuir, 1989, p. 47).

Scientists are as subject to self-deception and conflict of interest as anyone else is. Truth ultimately resides in the collective judgment of people who are committed to consensus and consistency. The social nature of science is clearly illustrated by the history of nonexistent phenomena.

My final case involves the A_2, an elementary particle created in experiments in which pi mesons from a high-energy accelerator collide with the proton nuclei of ordinary hydrogen. The direction and energy of the recoil proton after each collision are measured by a complex array of electronic detectors. The data from the detectors are stored on magnetic tape for later computer analysis. If a particle is created in some of the collisions, it appears as a bump in a plot of the analyzed data. All such bumps, technically called resonances, have similar shapes, which follow from fundamental theory. But in 1967, a group at the European Center for Nuclear Research (CERN) in Geneva found that the resonance of the A_2 had an anomalous dip in the center (Figure 8.2); the resonance was split (Chikovani et al., 1967).

This was a startling discovery, contrary to all experience. As seen in Figure 8.2, the CERN dip has only a few points in it, each with a probable error that is one-quarter the size of the dip. Still, the probability that this is a statistical fluctuation is less than 0.1 percent. Immediately, theorists started churning out papers to explain the greatest anomaly since parity violation.

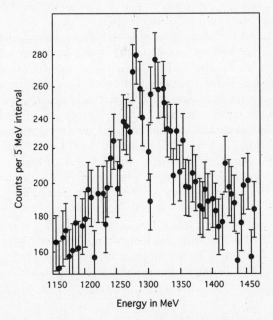

Figure 8.2 The A_2 resonance as reported by the experimental group from CERN (Chikovani et al., 1967). The anomalous (and erroneous) dip in the middle of the resonance was caused by a bias in the way the data were selected.

The position in energy of the A_2 dip had been so well determined by the CERN experiment that the experimental group from my department planned to use it to check an experiment they were doing at the Brookhaven National Laboratory on Long Island. To their great disappointment, they didn't see the dip. A dip is something that can easily be missed if there is a problem with the experiment, but something that is unlikely to be created by a problem. Thus the Northeastern group first thought that the problem was theirs, not CERN's. But after repeated checks of their equipment revealed no problems, and repeated experiments continued to show no dip, they announced their results at a stormy meeting of the American Physical Society in 1971 (Bowen et al., 1971).

A spokesman for the CERN experiment vigorously defended their result, claiming that the *fact* that CERN saw the dip *proved* that the CERN experiment had better resolution. To which Bernard Gottschalk, speaking for the Northeastern group replied, "Seeing spots before your eyes doesn't mean you have better vision." And so amid cheers and catcalls, the physicists argued their cases. Within a few months, the consensus swung to the Northeastern side, and the dip was never seen or heard of again.

This leaves the question of how a group of distinguished scientists, using the best equipment in the world, could see something that wasn't there. The CERN group, it seems, did exactly what Rhine had done—discarded data that didn't show what it wanted to find. As one CERN scientist so ingenuously explained to Gottschalk: "We broke the data into batches. Whenever we found a batch with no dip, we looked very carefully for something wrong and we always found something."[8] Because they didn't look so carefully for trouble when a batch showed a dip and because there is always something wrong in every run of a highly complex experiment, they managed to convert an initial statistical glitch into pathological science.

This story is worth telling because we are dealing here not with a few isolated individuals, but with a large team of highly trained scientists working with a mountain of electronic equipment and computers. As Langmuir commented about earlier nonexistent phenomena:

> These are cases where there is no dishonesty involved, but where people are tricked into false results by a lack of understanding about what human beings can do to themselves in the

way of being led astray by subjective effects, wishful thinking, or threshold interactions. These are examples of pathological science. These are things that attracted a great deal of attention. Usually hundreds of papers have been published upon them.

. . .

[But] the critics couldn't reproduce the effects. Only the supporters could do that. In the end, nothing was salvaged. Why should there be? There isn't anything there. There never was. (1989, pp. 43, 44)

Sophisticated equipment and training don't guarantee that subjectivity and self-deception won't creep into an experiment. Nor is there any guarantee that a substantial fraction of the scientific community may not, for a time, be taken in as well. But it is rare that the entire scientific community forms a sustained consensus based on marginal and unreproducible results. You can't fool all the people all the time.

In summary, then, scientific knowledge resides in the consensus of informed opinion. Reliable evidence and logical consistency are the two basic requirements for achieving such a consensus. These place a tight constraint on new knowledge, and scientists tend to be closed-minded about claims that aren't so constrained. This occasionally causes science to miss real knowledge for a time, as in the case of parity violation. But it helps guard against being swamped by a flood of nonexistent phenomena that result from the egocentric tendency toward self-deception. How wasteful this can be is shown by the case of cold fusion.

The fundamental conclusion of this book is that the social arrangement that is science is a new development in human history. It is, in fact, a sort of growing up, a putting aside of childish egocentrism and an acceptance of human responsibility in the face of human fallibility. But because science is so new, much of our thinking is still immature. We continue to seek a spiritual connection with the uncaring void, if not with the gods of the past, then with futuristic creatures from another planet. We can't accept that we may be alone in the universe. Are we? Does it matter? Chapter 9 takes up these and related questions.

9

Are We Alone?

The principal thesis of this book is that the development of scientific thinking on this planet was the result of particular historical conditions, not of inevitable natural forces. Natural forces permit the development of creatures such as us, but they don't mandate it. Natural selection, for instance, is purely opportunistic, responding to local conditions. It doesn't follow a script. Similarly, human biology didn't mandate the discovery of science. Science, too, was the result of a sequence of local factors.

Science has replaced the belief that we are the creation of a personal god with the idea that we are the product of natural forces. But we are also a product of history—that is, of innumerable local happenings and circumstances—which makes the whole process unpredictable and improbable. Many people who are comfortable with the idea that we are the creation of nature are disturbed by the thought that we also are creatures of chance. It seems preposterous to them that we are the only creatures in the galaxy that understand their origin. But in the earth's 4.5-billion-year history, this understanding came only in the last 150 years, hardly a less preposterous notion.

Technologically Advanced Extraterrestrial Civilizations

The existence of technologically advanced extraterrestrial civilizations (TAECs) have long been the mainstay of science fiction. In

H. G. Wells's *War of the Worlds,* the aliens are evil Martians, intent on conquest; in Fred Hoyle's *Black Cloud,* the alien is a superintelligence who has knowledge beyond earthly comprehension. Over the years, science fiction has gone from fearing extraterrestrials to deifying them. In addition, a variety of special-interest groups, from UFO cultists to scientists interested in the search for extraterrestrial intelligence, have so popularized the notion of TAECs that most people take their existence for granted. In one study, over half of the upper-class science undergraduates at a Canadian university thought that extraterrestrial spacecraft were currently visiting earth (Gray, 1987). Indeed, a universe densely populated with TAECs is the brave new cosmography of our time.

This high-density cosmography requires that the processes that lead from interstellar dust to advanced technology be prescriptive rather than historical. If the processes are more historical and less inevitable, the odds of creating a TAEC are smaller, and we are obliged to accept a lower-density cosmography. Although this seems to be a rather technical point that leads to only a quantitative change in the density, I have found intelligent people to get very emotional about it. As one engineer put it, "I don't want to think that we are alone in the universe." Having replaced God by nature, we want to think that nature knows what she's doing. The question of the density of TAECs may seem like the modern equivalent of the medieval question about the number of angels that can dance on the head of a pin. But there is more to these questions than meets the eye.

Although there isn't enough scientific evidence to decide the density question, it's worthwhile to outline how the question is usually approached. This will give us a language with which to discuss the issue. Once the issue has been clarified, I hope to show that the two cosmographies tell us essentially the same thing about our place in the universe. And that the question "Are we alone?" has a wonderfully satisfying answer.

Our sun is a star in the Milky Way galaxy, a vast system of some 100 billion stars. Astronomers estimate that there are at least 100 billion galaxies of similar size in the visible universe. The question of the number of TAECs in the universe naturally begins with the question of how many stars have planetary systems. We don't know this number, but we do know that perhaps 90 percent of all stars are in assemblages of two or more stars, which makes the maintenance of stable planetary orbits unlikely. Assuming, for the sake of argument, that 10 percent of the remaining solitary stars have planets, there would be a billion planetary systems in our galaxy.

How many of these planets have conditions suitable for life to originate? We can reasonably assume that for each planetary system there are only one or two planets at a distance from their star where the temperature is neither too cold nor too hot for life. In our system, these would be the Earth and Mars. As recently as 1975, when the *Viking 1* and *2* spacecrafts made soft landings on Mars, scientists could reasonably hope to find some form of primitive life there. But the *Viking* missions failed to detect life on Mars, dampening the optimists' view that life may be easily created under a variety of conditions. And *Voyager 2*'s spectacular images of Neptune and the other outer planets have demonstrated just how different from one another the planets are.

We know little about planet formation except that it is a very complex process, involving the segregation of the heavier elements (oxygen, silicon, iron, etc.) from the hydrogen and helium that constitutes over 98 percent of matter in the universe. So many factors are involved—mass, temperature, chemical composition, and so forth—that every planet will be different. Neither the earth nor any other planet in the solar system can be considered typical or even likely. Again, for the sake of argument, we will assume that 1 in 100 stars with planets has a planet with conditions suitable for the formation of life. Thus in the billion planetary systems in our galaxy, there are 10 million planets that already have life or will produce it.

Life on many of these planets would find extinction at an early stage because of mineral deficits or volcanic pollution. Life on many others would stagnate until extinction in a sea of blue-green algae, as earth did for a billion years. It was these algae that generated the oxygen in our atmosphere, a case of long-term pollution, since oxygen is toxic to all living organisms. Fortunately, the process was slow enough that some organisms had time to evolve enzymes to protect them from the oxygen and then metabolisms to exploit the more energetic reactions that the oxygen supported. The abundance of free iron in the early earth may have played an important role in slowing the oxygenating process, since it reacted with the oxygen as quickly as it was released. For a billion years, the earth literally rusted. The transition of a planet's atmosphere from oxygen-free to oxygen-rich is extremely delicate. If it happens too quickly, pollution could destroy all living things. If it happens too slowly, the planet's sun could burn out before higher life forms had time to evolve. Perhaps only 1 in 100 planets that developed life reached the stage of multicellular oxygen-breathing organisms. But that's still 100,000 planets in our galaxy that evolved organisms as complex as the sponge.

From this point, the main requirements for evolution to produce diverse new forms of increasing complexity are that the planetary environment be heterogeneous and moderately unstable. A homogeneous environment, such as a planet covered with water to a constant depth, wouldn't provide the variety of habitats to support a diversity of life forms. Similarly, a heterogeneous planet with an unchanging climate and geology wouldn't have the extinctions that provide the opportunities for new life forms. On the other hand, if a planet is too unstable, volcanic eruptions could release enough carbon monoxide and sulfur dioxide to poison all living things. Thus of the planets that reach the stage of multicellular organisms, perhaps 1 in 10 is heterogeneous and unstable enough to produce higher life forms but stable enough for life to survive for at least another 500 million years. This leaves us with 10,000 planets in the Milky Way that reached (or will reach) the stage at which the earth was 65 million years ago.

These numbers are, of course, just guesstimates, made for the purpose of discussion. However, it is similar to guesstimates made by others (Kaufmann, 1991; Shapley, 1963). Increasing or decreasing the number by a factor of ten won't affect our final conclusion.

But the next question is critical. Of the planets that developed higher life forms, how many developed higher intelligence? Before answering that, let's consider a less-charged question: Of the planets that developed higher life forms, how many developed flying creatures? This is easy. On earth, we know that insects, birds, and mammals (bats) independently evolved their ability to fly. So, assuming only a dense enough atmosphere, it's safe to say that flying is a likely outcome of evolution. However, the particular species of flying creatures that actually evolve depend critically on local effects. That is, although some form of flying creature is likely, no particular species is.

Higher intelligence, at the level of apes and dolphins—and perhaps some dinosaurs—also evolved independently several times. In analogy with flying, we can say that apelike intelligence will be a natural consequence of long-term evolution, but how it evolves further is a matter of local history. The hominid line in general, and our species in particular, is a consequence of local rather than global causes (see Chapter 3). So of the 10,000 planets that developed higher life forms, perhaps only 1,000 evolved creatures as smart as or smarter than we are.

For 200,000 years, we lived as unsophisticated hunters and cave painters. It was the exigencies of particular climatic changes that forced us on the road to civilization, another accident of history. This means that of the 1,000 planets that evolved humanoids, perhaps only 100 created civilization.

With civilization comes writing, literature, and philosophy. But as we have argued, not necessarily science. Too many local conditions seem to have been involved in the singular development of science on earth to consider it a very likely occurrence. Perhaps only 10 planets in our galaxy ever have developed or ever will develop our current level of technology.

The number 10 isn't to be taken seriously, of course. It could just as well be 100 or 1,000 or 1 or 0.1 without affecting my case. The important point is that the number is small in comparison with the immense size of the galaxy. A thousand technologically advanced civilizations scattered randomly throughout the galaxy would be, on the average, 2,000 light-years apart in space and 10 million years apart in time. That is, because each of these civilizations could have developed any time in the last 5 billion years, or in the next 5 billion years, our nearest neighbor may have burned up a 100 million years ago or may not develop for another billion years.

Thus it is very likely that at the present time ours is the only technologically advanced civilization in our galaxy. But because there are some 100 billion galaxies in the visible universe, we can be confident that there are a comparable number of contemporaneous technologically advanced civilizations, albeit separated from us by millions of light-years. Even if a few other TAECs currently exist in our galaxy, they are many thousands of light-years away, distances beyond the range of conventional methods of telecommunication,[1] let alone space travel.

This is the low-density cosmography. In it, we are the product of universal natural laws and peculiar local circumstances. This makes us rarer than many like to believe, but certainly not unique. Who is to say that one civilization per galaxy capable of understanding the grandeur of its own existence is too few?

The Search for Extraterrestrial Intelligence

John Ziman's definition of science—that it is the search for the widest possible consensus among competent researchers—is compatible with the historical view that science arose from the democratic impulses of the ancient Greeks. Science becomes pathological when it fails, for one reason or another, to honor this need for consensus. An isolated instance of such pathology can be a source of innocent merriment, as in the case of cold fusion, but should such pathology become widespread, it could endanger the entire scientific enterprise.

It may be difficult for a nonscientist to distinguish pathological science from real science. Even scientists can't always tell the difference. But there are telltale signs: Usually the subject under investigation is some marginally perceptible pattern that comes and goes in strange and arbitrary ways (see Chapter 8). With this perspective, let's examine the search for extraterrestrial intelligence (SETI), which is currently being funded at the rate of $10 million a year by the U.S. taxpayer (Kerr, 1992).

SETI advocates, such as Carl Sagan (1973), Frank Drake, and Thomas McDonough (1987), argue for a high-density cosmography with 10,000 to 100,000 contemporaneous TAECs per galaxy. But even 100,000 TAECs would be relatively rare, only 1 in every million stars, and the average distance between them would be more than 400 light-years, still beyond the range of telecommunication and space travel. Even if, by chance, there were a TAEC on nearby Barnard's star, only 6 light-years from earth, it would probably be beyond the range of intelligible communication and certainly beyond the range of space travel by any means known to science.

For example, signals broadcast by *Voyager 1* are still being detected, even though the spacecraft is 5 billion miles from earth (Wilford, 1993). But Barnard's star is 7,000 times farther away, so a TAEC there would need 50 million times more transmitter power to send the same size signal to earth. A TAEC 100 light-years away would require 12 billion times more transmitter power.[2]

Although such monster transmitters aren't impossible, it's doubtful any TAEC would build one just to signal us of its existence. Frank Drake, a long-time SETI advocate, sees it differently: "You can theorize forever. But intelligent life is so complicated in its activities and philosophies, as we know from ourselves, that it's quite impossible to psych out the extraterrestrials and deduce by logic how they might behave. The only way we can really learn the truth is to search" (quoted in Broad, 1990, p. C1).

This seems scientific and disarmingly undogmatic, but it ignores the fact that it is impossible to prove something doesn't exist. No amount of negative results will ever produce a consensus that TAECs don't exist. Nor will a positive result produce a consensus that they do. The signals received by radio telescopes are a complex jumble of electromagnetic radiations from natural and artificial sources. Any alleged TAEC communication will have to be teased out of all this static by sophisticated computer analysis, and it is unlikely ever to be unambiguous enough to convince everyone. Independent scientists analyzing the same data will reach different conclusions. SETI scientists will claim this is because the outside scientists' methods aren't as clever as theirs. Congress will be besieged for more money. Critics will be accused of wanting to hang up on E.T. (McDonough, 1987). Emotions will run high as the level of discussion deteriorates.

The marginal and ambiguous nature of SETI puts it in danger of becoming pathological (Tarter, 1993). Because non-SETI scientists won't have SETI's highly complex and specialized analytic techniques, a situation could arise in which the normal checks and cross-checks of science are impossible.

Does this mean that the search for extraterrestrial life should be abandoned? The direct search, yes, as I believe it can produce only ambiguous results that will weaken the scientific enterprise. But as our earlier discussion showed, many aspects of basic science pertain to the question of our origins. The more we learn about the formation of planets, the origin of life, and the evolution of our species, the better we will be able to estimate the likelihood of there being other TAECs in the galaxy. In this way, we can make progress in our understanding of our place in the universe while staying within the scientific consensus. Already we can be reasonably certain that there are between 1 and 100,000 TAECs in our galaxy and that the nearest TAEC is likely to be forever beyond our reach.

Of course, forever is a long time. Isn't it possible that some TAEC, which may have had technology for millions of years, has discovered practical methods of interstellar travel and communication? The SETI advocates think so. Indeed, much of their program is based on the premise that there may be a TAEC out there that can do things we can't imagine. I disagree, for reasons I will explain in the next section. If I am right—if it isn't possible ever to communicate with beings on the nearest star or even to know of their existence—then it doesn't matter what the density of TAECs is. Whatever the case, we are, for all practical purposes, isolated on this planet.

Why is this thought so frightening? We never had any reason to think otherwise. I believe the answer lies in our peculiar penchant for egocentric thinking, our need to feel connected to the forces of nature. Somehow we feel more connected if we think that our existence is the consequence of universal principles and not merely of local circumstances. Space sagas take on a mythic dimension when we realize that they are based on the idea that we are part of a larger pattern of existence.

Comforting as this idea is, it's just Bronze Age egocentrism in a space suit. Many cults have already arisen in which extraterrestrial beings have replaced earthly prophets as the bearers of ultimate truth. One example is *The Urantia Book*, a 2,097-page tome published in Chicago that claims to be the work of extraterrestrials channeled through an unknown earthling (Gardner, 1990). Science is a weak defense against such nonsense, but it's the only defense we have. SETI advocates, because of their zealous enthusiasm for extraterrestrial communications, give scientific credence to myriad space fantasies.

It is time that enlightened human beings accept the truth about their existence: Our thoughts and wishes have no connection to the vast universe that surrounds us. This universe wasn't created for our benefit, nor we for its. We are forever trapped on our small blue planet, made ever smaller by our advances in transportation and communication. We are the supreme beings in our part of the galaxy and thus have the ultimate responsibility for what happens here. There are no superaliens to rescue us.

Progress

Our view of progress is intimately tied to our cosmography. For if progress is unlimited, then we are free to believe anything, even the possibility that superaliens have solved the problem of interstellar travel. Carl Sagan, speculating on TAECs hundreds of millions of years old, stated that they "will have discovered laws of nature and invented technologies whose applications will appear to us indistinguishable from magic" (1973, p. 366). We would appear to them as backward as our Paleolithic ancestors appear to us.

This, I believe, expresses the common ideas that progress is a steady accumulation of knowledge over time and that knowledge is unlimited. But the discovery of the external world and, with it, the rejection of magic have changed forever the way we think. We are hardly likely to mistake extraterrestrials for gods, no matter how advanced their technology. Furthermore, the idea that knowledge is unlimited is at odds with the belief in an external world. For if there is a world apart from our thoughts and if this world operates according to a set of consistent rules, then once we know these rules, we know the basic mechanisms that govern the world.

Inside the atoms are nuclei; inside the nuclei are protons and neutrons; inside the protons and neutrons are quarks. It's conceivable that inside the quarks are still deeper particles and that inside these are still more, forming an infinite sequence that can never be fully known. But it is the proton and neutron that govern the world on a human, and even galactic, scale, and these we understand in great detail. All future technological advances will be made with the atoms and forces we already know. Although these can be arranged in endless combinations, there are limits to what is possible.

The explosive growth in the speed of transportation in the last 200 years leads naturally to the belief that still greater speeds will occur in the future. Yet it is probable that human beings will never travel significantly faster than did the *Apollo* astronauts as they left the earth for the moon (at about 25,000 miles per hour) and that normal passenger travel will never exceed a few hundred miles per hour on the ground and a few thousand miles per hour in the air. Indeed, ground speeds of 100 miles per hour are only a little more common today than they were 100 years ago.[3]

The notion of unlimited technological advancement is so ingrained in our worldview that laypersons and scientists alike consider any assertion of a limit to the possible to be unscientifically dogmatic. We all know stories of prognosticators whose negative claims that people would never land on the moon or that computers would never be in the home were proved wrong by events. We forget that there are other negative claims, such as the impossibility of building a perpetual-motion machine and the impossibility of traveling faster than the speed of light, that have never been successfully challenged.

Limits, it seems to me, are inherent in science itself or, rather, in the belief in an external world on which science is based. Living, as we are, in the adolescence of science, we are still dazzled by its accomplishments and unmindful of these limits. But as science matures, these limits will become more apparent and will have a profound effect on our attitudes.

Consider my statement that we have already reached the limit of speed for a human being of about 5 to 10 miles per second (18,000 to 36,000 mph). At these speeds, men have traveled to the moon in a few days and could travel to Mars in nine months. But beyond that, there is nowhere to go. The nearest star is so far away that it would take a spaceship traveling at 10 miles per second 80,000 years to get there. Of course, scientists have thought hard about exotic propulsion systems that could make interstellar travel possible (McDonough, 1987). But nothing remotely practical has come of this (Simmons and McInnes, 1993). Because all matter is made of a few kinds of atoms, there are severe energy and mass restrictions on the design of any futuristic spaceship.

Barnard's star is the star closest to earth that could maintain a stable planetary system.[4] Indeed, there is some evidence that it may have at least one Jupiter-size planet. So let us consider what it would take for Barnardians to travel to earth.

Earth is 6 light-years, or 33 trillion (33,000,000,000,000) miles, from Barnard's star. It is just a matter of arithmetic to calculate the speed that a Barnardian spaceship needs to reach the earth in a given time. For example, if the journey is to take 60 years, the speed must be 1/10 the speed of light, or 18,600 miles per second; for a 600-year trip, the speed must be 1/100 the speed of light, or 1,860 miles per second; and for a 6,000-year trip, the speed must be 1/1,000 the speed of light, a leisurely 186 miles per second.

The most fundamental law in all of physics—the conservation of momentum—tells us that the only way a spaceship can increase or decrease its speed in the vacuum of space is by blasting out matter at high speed. Let us assume that the Barnardians are patient creatures, willing to travel for 6,000 years in order to make the headlines of the supermarket tabloids. In order to make a soft landing on earth, their spaceship must slow down from 186 miles per second to the earth's speed around the sun, about 19 miles per second, by firing its engines in reverse.

The amount of matter that the spacecraft needs to expel depends on its weight and the exhaust speed of the rocket propellant. For example, a fifty-ton spacecraft—the size of the one that carried the *Apollo* astronauts to the moon—would have to expel more than 50 trillion tons of matter at a speed of 6 miles per second (twice that of any known propellant). This means that a small fifty-ton Barnardian spaceship would have to reach our solar system attached to a fuel tank 3 miles wide and 1,000 miles long to be able to slow down to earth's speed. To have launched such a monster, the Barnardians would have needed a rocket larger than their planet. All this to send a spacecraft—only large enough to support three human beings for a week—on a one-way, 6,000-year journey to earth. Hardly something you would expect intelligent beings to do.

The fundamental difficulty with space travel is that a spaceship must carry with it all the fuel it will ever need, to both increase and decrease its speed. There are no filling stations in space. But the bigger the spaceship's fuel tank is, the larger the total mass that must be accelerated. It's a vicious circle. More fuel is needed to go faster, but the increased fuel adds extra weight to the rocket, making it more difficult to go faster. As a consequence, with larger fuel tanks, there is painfully little increase in the final speed.[5]

Can we improve matters by imagining some exotic futuristic fuels? No, because the critical characteristic of a rocket propellant is how fast it leaves the engine. In a chemical rocket, this depends on the temperature at which the propellant burns and how light it is per atom. There is a highest temperature—the melting point of metal—and a lightest atom—hydrogen. This puts an absolute limit on the exhaust speed of a chemical rocket that isn't much greater than what has already been achieved on earth.

To get around the limitations of the chemical rocket, futurists like to dream about an ion-propulsion engine. This would contain a particle accelerator capable of shooting out hydrogen ions at speeds approaching the speed of light, 186,000 miles per second. Since this is much greater than the few-miles-per-second ejection speed of a chemical rocket, much less propellant would be needed. Indeed, one could imagine an ion starship about the size of a nuclear submarine. Its nuclear reactor would run an electric generator that would power a particle accelerator—all known technologies. Such a compact starship might indeed be capable of reaching speeds greatly in excess of those of a monstrous chemical rocket. The only drawback with ion engines—and the reason we don't build them—is that they gain speed very slowly.

On a journey from Barnard's star to earth, an ion-propelled starship would have to accelerate for thousands of years to reach the speed of a chemical rocket. By that time, it would be halfway to earth and would have to reverse its engine's thrust, taking several thousand more years to slow down to earth's orbital speed. Throughout its 5,000- to 10,000-year journey, all the ship's machinery would have to function flawlessly.

The idea that spaceships may someday trek from star to star makes great science fiction but bad science. The laws of physics and the properties of matter limit the speed of spaceships, making it impossible to travel between stars in any reasonable time.

This is a surprisingly unpopular message. Instead of being relieved that we can't be invaded by hostile aliens, people are disturbed by the idea that our technology is approaching some theoretical limit so soon after it first started. Only 160 years separate the first steam locomotive from manned space flight, a stupendous rate of progress. Surely if technology could take us from 3 miles per hour to 25,000 miles per hour (7 miles per second) in only 160 years, someday speeds in excess of 186 miles per second should be possible.

Unfortunately, technology doesn't work this way. Human experience shows us that new technologies reach their practical limits early in their development. The Great Pyramid of Egypt was built about 2680 B.C., only a few hundred years after the Egyptians first learned to cut and move large blocks of stone. They continued building pyramids for another 1,000 years, but never again one as big.

Men landed on the moon only eight years after Yuri Gagarin's first orbital flight. The moon landing may be our Great Pyramid, an accomplishment never again to be equalled. Although a manned mission to Mars may be technologically possible, it's questionable whether it could be justified by its costs and risks. Even if peace and democracy should reign throughout the world and $1 trillion a year in armament expense is no longer needed, there will be many other things to do with the money.

None of this is to imply that we are running out of things to learn. But clearly, new knowledge will come primarily from areas in which we are relatively ignorant, not from mature areas in which our knowledge is essentially complete. There is no light-weight element or law of motion yet to be discovered. It is from the fields of molecular biology, brain research, and computer technology that the epochal discoveries of tomorrow will come.

The twenty-first century may see major breakthrough in such areas as thinking, aging, and behavior. In areas in which we are still ignorant, we can freely speculate about processes that increase our intelligence or medicines that retard or eliminate aging or cures for violent and destructive behavior. As we learn more about these subjects, we will come to understand what is possible and what isn't. And remember, even a Homeric god couldn't do the impossible.

The Search for Terrestrial Intelligence

I have tried to avoid the word *intelligence* as much as possible in this book, because of its confusing ambiguity. It means one thing in the phrase "extraterrestrial intelligence" and quite another in the phrase "animal intelligence." In the first phrase, intelligence refers specifically to the ability to analyze highly technical data using sophisticated mathematical techniques, and in the second phrase, it refers generally to the ability to acquire, analyze, and act on information about the external world. In the first sense, intelligence is possessed by only a small minority of the human population, whereas in the latter sense, intelligence is possessed by all living creatures.

A bee can find a nectar-bearing patch of flowers from the polarization of skylight and information communicated to it in the hive. A bat can detect and capture a flying moth in complete darkness, using echo-locating high-frequency sound. A bird can navigate over thousands of miles of open water, using the position of the sun and stars and the earth's magnetic field. It would surely be egotistic of us to deny the term *intelligent* to these creatures while retaining it for a creature that until quite recently built its wells in back of its outhouses.

Of course, most animal intelligence is instinctive and specialized. Bees don't go to flight school; they appear to be born with all the abilities they need. It's all hardware with them. For us poor creatures, born with a blank slate, everything must be painfully learned. It's all software with us.

We are, to a large extent, an evolutionary oddity: a social animal bound together not by the immutable laws of genetics, but by the rules and customs of our own making. In this position, we can look godlike one moment and ridiculous the next. At our best, we can come together in great international congresses to contract peace among nations and to provide relief to refugees on the other side of the planet. At our worst, we can kill our own brother for the change in his pocket.

Intelligent isn't the proper word for us; perhaps *educable* would be better. As educable creatures we are without peer on this planet and perhaps in this galaxy. As intelligent creatures we are one among several million other species on earth, most of which have abilities beyond our current understanding. In this sense, we are hardly alone, with millions of fellow species to delight, inspire, and educate us.

Education is the hallmark of humankind. It goes on all the time, in the home, in the street, and in the school. No community is without it. But modern civilization is unique in requiring a formal "Greek-style" education for everyone. The results have been spectacular. In two centuries, reading and writing, once the possession of the few, have become the right of all. Still, expectations for education have outpaced performance, and in the United States, public education is in crisis.

The problems with education in general and science education in particular are discussed in Chapter 10. This final Chapter describes some current initiatives to improve the quality of science education and ends with an audacious proposal for the salvation of the U.S. public-school system.

10

Education for an Age of Science

Science is a social activity that studies those things for which a universal consensus is possible (see Chapter 8). Its methods are experimentation and mathematics because it is possible to obtain general agreement using them. They are, to use Ziman's (1968) term, *consensible*. This isn't because these methods are unambiguous—the case of cold fusion shows just how ambiguous experiments can be— but because their ambiguities are capable of resolution.

There are procedural and analytical difficulties in even the simplest experiments. For example, to determine the period of a pendulum—the time it takes it to make one back-and-forth swing—care must be taken with how the string is attached to its support and how the time for one swing is measured. Most students don't use the best procedure when first doing this experiment, but in time they come to understand why there is less error if they measure the time for twenty swings (and divide it by twenty) than to measure the time for only one swing. To disentangle the effects that the mass and the length of the pendulum have on its period, the student must learn to control several variables.

The pendulum experiment is an excellent example of the modes of thought associated with science. Many of these modes, such as the control of variables, are characteristic of the formal operations stage of intellectual development. In Piaget's (1929/1960) theory of mental developments, this stage emerges only after a lengthy period of concrete thinking. Educational research over the last twenty years has confirmed these essential features of Piaget's theory and added the important insight that the formal operations needed for scientific and logical thinking require specific prerequisite experiences. Many studies have shown that more than half of adult Americans never reach the stage of formal operations (Arons and Karplus, 1976), meaning that they can't analyze a situation with several variables or understand a simple syllogism.[1]

From these studies and from my own experiences with students of all ages, I have derived the central assumptions of this book: that scientific thought is very different from ordinary thought and that historically scientific thinking could have developed only in those societies that provided the experiences that are prerequisite to it. This gives science a unique history, as only Greece, and cultures influenced by Greece, had these prerequisites. Today, of course, science has a place in all cultures, but individuals still must have the proper sequence of experiences in order to develop scientific thinking.

Teaching Science

Among a science teacher's most striking experiences are encounters with bright, eager students who are utterly unable to understand some seemingly simple scientific idea. The intelligence of these students is apparent from the clarity with which they recognize their lack of understanding. Nevertheless, they are unable to process the information they are given because they are missing some critical mental structures.

These structures develop from a sequence of relevant experiences that extends back to earliest childhood. In simpler times, most children of the same class and culture had similar experiences and so progressed in roughly the same direction. But now, the immense variety of toys and activities available to children provide a multitude of different experiences, few of which, unfortunately, are relevant to the development of scientific thinking. One peek into the room of a middle-class six-year-old girl reveals a world of stuffed toys and soft fabrics. There are no toys that fit together, nothing with a straight line. By age four or five, the child may have lost all interest in the type of toys that supply the manipulative experiences relevant to developing the mental structures necessary for mathematics.

Contrast this with a three-year-old who plays with toy trains.

At first, the child may play randomly with the individual parts but, if guided by a parent, will soon become interested in connecting them "properly." The train soon imposes its own reality, and the child comes to understand that there are right and wrong ways to run a railroad. Playing with trains comes to mean connecting tracks so that cars can roll on them from one place to another. From such play comes the mental structures associated with lines and connections, which enable the child to see the geometrical similarity between a real train on real tracks and a toy train on toy tracks. Some such insight seems to be necessary for the development of the formal concepts of scale and proportionality.

Students who reach middle school without the relevant concrete experiences with similarity aren't prepared for subjects that involve proportionality, such as percent and time–rate–distance. The failure to understand these subjects blocks any further progress in mathematics and science. To prevent this blockage, students from kindergarten on must be provided with concrete activities designed to develop geometric and quantitative thinking.

I have stressed throughout this book that egocentrism, the innate human tendency to confuse thought and reality, is a natural barrier to the development of science. It is also the principal obstacle to the learning of science and scientific thinking. Piaget (1929/1960) pointed out that an early manifestation of egocentrism is the inability of children as old as nine or ten to distinguish words from the things they represent. This confusion is perpetuated by an educational system that teaches in terms of words rather than experiences.

Just as children need experiences constructing things with rigid objects in order to develop a sense of quantity and proportion, they need concrete experiences with an externally imposed reality to grow away from excessive egocentrism. In order to develop a sense of logic, these experiences must involve outcomes that are recognizably right or wrong. As Philip and Phylis Morrison put it: "The currency of what is right and wrong and true and false is not a currency of word alone, but also depends on citation of relevant experience" (1989, p. 5).

Words name aspects of our experiences. Without the experiences, the words are meaningless. What does *green* mean to a blind person? Or *abstinence* to a teenager?[2] Most of students' ambiguous, inconsistent, and illogical uses of words are not corrected, thus further perpetuating their egocentric confusion of thought and reality, of words with their meanings. Laboratory activities are needed to provide the experiences for which words like *circuit* and *switch* are meaningful and in which inconsistency and illogic are self-evident. As the Morrisons observed: "The teacher doesn't have to tell you the circuit you built is not working. You know it's not working because it doesn't light the light. And when it *does* work, you know that too. You begin to form a sense of evidence and self-reliance" (1989, p. 5). And, I would add, a sense of reality.

Every good teacher knows it's better for students to do than to listen. When teaching subjects such as optics, electricity, and cellular physiology, it is imperative. Elementary- and middle-school students should have done experiments with light and circuits and made observations of cells with a microscope before they encounter much of the technical vocabulary. The words should arise naturally as the names for objects and relations already experienced.

Contrast this with the following scenario that elementary-school teachers are given on how to teach the abstract concept of gravity to first and second graders:

> Mr. Jones tells the children that the thing causing his pencil and their crayons to fall is called *gravity*. Gravity cannot be seen, smelled, heard, tasted or touched—but it is real. Children can see the effect of gravity on the pencil and crayon as well as other objects they may test. (Zeitler and Barufaldi, 1988, pp. 29–30)

The confusion in this scenario of the name of an explanation for an explanation is classic nonscientific thinking. It is pervasive throughout elementary- and middle-school science because the teachers, and the teachers who teach them, are nonscientific thinkers. Many students leave school thinking that Newton invented the word *gravity* and wondering why it's such a big deal.

Newton's law of universal gravity isn't a name, but a relationship. It relates, for example, the falling of an apple or a crayon to the motion of the moon around the earth and the motion of the earth around the sun (see Chapter 7). Of course, the study of this relationship is beyond elementary-school students, but the study of falling is not.

The key to the scientific investigation of falling, or any other phenomenon, is to focus on the phenomenon. The basic observation is that objects fall when released. For elementary students, gravity is just the name of this phenomenon. As with any name, it doesn't tell us anything we didn't already know.

To get more information about falling, the second graders might be asked: "Will a large piece of clay fall faster than a small piece of clay?" After some discussion, they could be asked to find out for themselves and to test a rock and a lump of clay. Then there is the procedural problem of releasing two objects at the same time. How can you improve on this? Can you make things fall slower? Or faster? The subject of falling is rich enough to occupy many days of investigation without ever going outside the scope of the students' own experience. Older students can continue the investigation measuring time with stopwatches, rolling spheres down inclines, and making parachutes.

But if the teaching of science is as easy as this, why isn't it being done? This is a critical question with a complex answer. In the first place, this approach to science isn't as easy as I make it sound. The questions I suggest are carefully chosen for their simplicity and relevance. I don't, for example, suggest that students compare the motion of an air-filled balloon with that of a helium-filled balloon, because such a comparison, though extremely interesting, is neither relevant nor simple. Elementary-school teachers have a legitimate fear of science because it's so easy for them to get in over their heads.

In order to teach science phenomenologically, the teacher must have learned science in this way, but very few have. Indeed, most elementary-school teachers have never had a college-level science course, and those who have are only a little better off, because college courses suffer from the same problems of over abstraction that we want to overcome.

In the second place, there is no general agreement among educators on how science should be taught. Although most would agree that some activity-based learning is essential, there is wide disagreement about how structured these activities should be. Some reformers strongly advocate a discovery approach in which labs are open-ended (Yager, 1991), whereas others (including myself) prefer a highly structured approach in which students are guided past the many pitfalls in the way of a successful experiment. Behind this seemingly minor technical difference are major differences in the assumptions of what science is and how much it differs from ordinary thought.

In the third place, conflicting agendas are often involved in the development of new curricula (Beardsley, 1992). This has unfortunately been the case with Project 2061, a science education initiative of the American Association for the Advancement of Science (AAAS).

The AAAS's 135,000 members come from both the natural and the social sciences. Consequently, the Project 2061 report *Science for All Americans* makes recommendations that range from decrying the fate of Native Americans to explaining the principles of engineering design (American Association, 1989). Clearly, nothing worthwhile will happen if science education tries to fulfill a multitude of agendas.[3]

Scope, Sequence, and Coordination (SS&C) is a project of the National Science Teachers Association (NSTA) to develop a coordinated sequence of hands-on experiences that leads to successively higher levels of abstraction (Aldridge, 1992; National Science Teachers Association, 1992). As Aldridge originally conceived it, students would study a topic, such as the pendulum, a number of times during their school years. Each time, the topic would be treated with the sophistication appropriate to the students' current level of intellectual development. Specific curricula are now being written by groups in five states and Puerto Rico, funded by the National Science Foundation (NSF). Although such national projects tend to lose focus as many players with many different agendas become involved, if SS&C stays fixed on its original concept, it could become an effective national model.

The aim of science education should be to supply to students the prerequisite experiences necessary to develop scientific thinking. This is a complex task, made even more difficult by the lack of any general agreement on what science is. But if we accept the view of this book, that science is the search for consensible knowledge, then its basis is the convincing observation, experiment, and deduction. Critical thinking is needed for this in order to recognize sources of errors in experimental procedures or logical deductions.

Other tasks, such as forming hypotheses and recognizing prejudices, aren't basic to science (Storey and Carter, 1992). Asking a student to make a hypothesis about a phenomenon she has just seen for the first time only supports her egocentric belief that she has inside her a hidden source of knowledge about the external world. Asking a class for its opinion about which way an experiment will turn out is good for getting its attention and to show that class opinion isn't a valid source of knowledge. But serious class discussion should be limited to matters that are within the class's range of knowledge: What were the major sources of error in the last experiment? What can be done to decrease these errors?

Teachers teach with words because that's how they were taught. But in science, the thing is the thing. Elementary-school students, being concrete thinkers, love physical-science experiments. Mirrors alone provide endless opportunities for investigating diverse phenomena and developing critical thinking. The object isn't to give these phenomena fancy names or to explain them in terms of more advanced theories, but to develop consistent rules that relate one mirror experiment to another. This is the only way that a concrete thinker can learn about logical consistency and mathematical relationships and so advance toward more formal modes of thought.

Teaching the Teachers

Overshadowing all attempts at curricular reform is the lack of knowledge of science and mathematics of most American teachers. Almost all—including the best middle-school science teachers—are very weak in physical science. Mathematics is seldom used in either elementary- or middle-school science, let alone history or social science, even when the same teacher is teaching both subjects to the same students.

Imagine if reading were taught forty minutes a day in reading class, but students and teachers never read anywhere else. How many students would learn to read? But this is exactly how mathematics is taught. Science, history, and social studies don't use mathematics on a regular basis, so students have no idea how it can be used to make sense of their lives. What percentage of your diet is fat? If Americans buy 300 million compact discs a year, how many pounds of trash can be saved by not packaging them in long wrappers? If the budget deficit of the United States is $400 billion, what is the deficit per capita? The average nonscience college student can't do problems like these.

This is a self-perpetuating problem, since it is the nonscience and nonmathematics students who go on to become teachers. Furthermore, education schools have long been isolated from the rest of the academic community. Setting their own curriculum and standards, they have seldom involved scientists in the design of science education courses for their students. Science education has become its own specialty, training its own practitioners to teach others how to teach science.

There is a desperate need to enhance teachers' fundamental knowledge of science and of the teaching of science. This can't be done by conventional college-level science courses, which are too abstract for the education students and their future students. Science must be taught to teachers using the same concrete methods that they are to use with their students.

Since 1990, my colleagues and I have conducted in-service enhancement programs for middle-school science teachers through Project SEED (Science Education through Experiments and Demonstrations).[4] The core of these programs is the SEED Sourcebook of 150 demonstrations, activities, and experiments covering the basic physical concepts of length, time, mass, volume, force, pressure, work, speed, acceleration, elements, compounds, solar system, temperature, heat, sound, light, electricity, and magnetism (Cromer and Zahopoulos, 1993). Each summer a selected group of middle-school science teachers covers this material in thirteen day-long workshops.

Following Piagetian principles, the course starts with the simplest concepts and builds up in complexity through a sequence of concrete activities. It is assumed that many of the teachers in our program are, like their students, largely at a concrete operational level of cognitive development. Progress toward formal operational thinking is promoted through inquiry-based activities and the use of simple mathematical techniques (ratios, averages, graphs) to analyze quantitative data. The course challenges the best teachers and provides some useful ideas for the weakest.

The rules for successful science teaching are well known (Padilla, 1983): Be concrete, be active, be quantitative. It is the translation of these rules into a coherent program that is challenging. We begin the course with an exercise in measurement. Each teacher cuts out a rectangular strip of poster board and subdivides it into eighths by folding it three times. The strip is then used to measure the length and width of a table. The results are put on the board. Of course, each teacher gets a different set of numbers because each is using a different length of poster board. As obvious as this is, just think of trying to explain the concept in words. What is clear in practice is almost impossible for a sixth grader to understand in the abstract.

Next, each teacher divides her own measurement of the table's width by her measurement of its length. Surprisingly, most of the teachers get about the same value. (Values very different from the majority's show the prevalence of procedural errors and the need to repeat measurements.) The width of a rectangle divided by its length is an intrinsic property of the rectangle, its rectangularness. The teachers then cut out of the poster board several rectangles of different sizes but with the same rectangularness as the table. This requires some more mathematics. Finally, these rectangles are inserted in clay feet and arranged upright on a table so that when viewed from one point they all appear to be the same size.

The purpose of these activities is to provide concrete experiences with such abstract concepts as ratio, proportion, and similarity. Historically, the concept of a measure was basic to Greek mathematics (see Chapter 5), and developmentally, experience with measurement is a prerequisite for understanding proportion and similarity.

It may be surprising that measuring length is something that must be taught in school. Of course, many children learn this at home, but many don't. Inner-city teachers modified the measurement lesson by having their students take their strips home to measure things around the house. One girl got so excited at her success at this that her other course work started to improve: She had learned that she could learn (Leventman, 1992).

Along with the SEED enhancement program, we have a Leadership Institute for SEED alumni. The institute trains Leader Teachers to teach afterschool workshops to teachers in their school districts. Often referred to as "teachers teaching teachers," it is a mechanism for reaching a much larger teacher population than can be reached by summer programs. During the summer Leadership program, the Leader Teachers develop and conduct two-hour workshops for one another, and decide on six that they all will offer to teachers in their school districts during the academic year.

Via satellite television and telephone, my colleague Christos Zahopoulos and I communicate six times a year with the Leader Teachers and their workshop participants to discuss issues and exchange information. Above all, we want to break down the isolation of the classroom teacher by creating a community of dedicated professionals.

Ambitious as this program may sound, it provides only twelve hours of instruction a year to each participating teacher, less than a quarter of the instruction time provided by the summer SEED program. Only if it is sustained over many years, with workshop attendance a regular part of a teacher's professional activities, can SEED make even a modest improvement in the quality of science education.

Supporting the Teachers

Activity-based teaching is very demanding. Besides designing appropriate activities, report sheets, and tests, the teacher must maintain a stock of materials, respond to challenging questions, and keep law and order. Here school volunteers can be invaluable in supplying in-class support. Many school districts have well-developed volunteer programs. Ideally, the science volunteer should be someone with a technical or scientific background.

Through support from the National Science Foundation, we have established RE-SEED (Retirees Enhancing Science Education through Experiments and Demonstrations) which provides SEED training to retired engineers and scientists interested in volunteering their services to science teachers. The retirees are trained in twelve day-long sessions in April–May and September–October, and in September, they start assisting a middle-school science teacher one day a week. Now in its third year, RE-SEED has thirty of these very special people bringing their expertise and experience directly into the classroom. They are important agents for change in the schools, as they can make that vital connection between mathematics and science that teachers find so elusive.

The SEED training is essential to the volunteers, since it orients them toward the basic concepts and elementary experiments of middle-school science. Many of the volunteers have recently retired from high-tech engineering jobs and are delighted to review the fundamental principles that underlie all of science. With support from the Noyce Foundation, we have started to train retirees throughout Massachusetts to be RE-SEED trainers in their own communities. If retirees can successfully teach the SEED material to other retirees, a nationwide corps of specialized volunteers could be created in a few years.

Training New Teachers

In-service training and in-class assistance are ways of immediately improving science education. In the long run, however, we must improve the science training of students going into teaching. Traditionally, the education schools have resisted forcing their students to take demanding science courses, and the science departments haven't wanted to teach sissy science courses.

The problem is that there is no such subject as "science" at the university level. You can study physics, chemistry, biology, or even science education, but you can't study science. Introductory courses in physics, chemistry, and biology are aimed at qualifying the students for more advanced courses in their respective fields. There is an enormous gap between what these courses teach and what a teacher needs to know to teach science at the elementary- or middle-school level.

Is there such a subject as science? I believe there is. The unity of science is an important theme of this book (see Chapter 1). It is widely recognized by educators, who are calling for a more integrated approach to the teaching of science. However, integrated curricula present problems of design and execution. For two years (1991–1993), I taught my own integrated science course to nonscience college students. My experience with this course is given in Appendix B, along with a typical final examination. The course tried to piece together into a meaningful pattern the students' shards of knowledge of science. This was possible for only a few, since most didn't have an eighth-grade knowledge of science or mathematics.

There is nothing surprising in this. Students who haven't formed the mental structures for abstract thinking by eighth grade aren't likely to develop them later. The educational system accommodates them, rather than educating them. My course used a textbook (Krauskopf and Beiser, 1991) that is less demanding than some ninth-grade physical science textbooks (Carle, Sarquis, and Nolan, 1991; Hurd et al., 1991). This isn't meant as criticism of these books. These are sound books, suitable for a ninth grader (or a college student) who can handle proportions and control variables. I would be elated if nonscience college graduates knew half of what was in any one of them. The problem is that even by ninth grade, the textbooks are too formal for the majority of students who are still concrete thinkers (Lawson and Renner, 1974).

What is needed is a totally different type of course that approaches science much more concretely. At Brooklyn College, Michael Sobel is developing a laboratory-based physical science course that teaches science to education students in the concrete way in which they should teach it to their students. Sobel, a physics professor and my long-time hiking companion, is part of the School of Education's NSF-sponsored project to improve the teaching of science to education students. The project's involvement of both education and science faculty is a commendable breakdown of the barrier that usually separates these groups.

Sobel's course uses activities and experiments from the SEED Sourcebook and elementary- and middle-school textbooks to teach the basic physical concepts. In this way, the students learn both the concepts and how they can be taught to their students. But isn't this how teachers are normally taught? Unfortunately, it isn't. A standard textbook in elementary science education (Zeitler and Barufaldi, 1988) isn't about science. It covers the history of elementary science education, educational theories, availability of resources, designing a teaching unit, and so on, but the education student is expected to have learned the science elsewhere. Since elsewhere doesn't offer the courses teachers need, teachers remain untrained in science education.

The Full Option Science System (FOSS), developed by the Lawrence Hall of Science (University of California at Berkeley), is an elementary science program, complete with workbooks, activities, and teacher guides. It offers a shelf of material, more than enough for a college-level course in science. Scientists should be teaching this material to elementary-education students. It isn't glamorous work, and the NSF probably won't give anyone a grant to do it, but I can't think of a quicker or better way to give the future teachers of America the training they need to do their job.

The Crisis in American Education

The crisis in science education in the United States is only one aspect of the general crisis in education. American students are behind students of competitive countries not only in mathematics and science but also in history, geography, and foreign languages. My nonscience students had a harder time with the history in my course than with the mathematics, repeatedly putting Muhammad before Moses on tests despite all I said to the contrary (see Appendix B). In a future in which prosperity will depend more and more on a knowledge of science and an understanding of other cultures, our population is rapidly falling behind. Already, American industry is unable to use the most advanced manufacturing robots because its workers don't have the reading and mathematical skills necessary to operate them. Japanese workers do. Such machines increase worker productivity, and hence the national standard of living, by relying on intelligent workers rather than engineers to reconfigure them as needed.

The American crisis in education is fundamentally one of values. Americans value childhood for its own sake, not primarily as a period of preparation for adult life. Consequently, education is permitted to compete with many other activities—ballet lessons, sports, part-time work, television, dating—for a student's time. Our students don't go to school on Saturday; Japanese students do. Half our students don't have desks at home where they can do their homework; almost all Japanese students do. Most American teenagers do some paid work during the school year; most Japanese don't. Our students attend school in torn jeans and Megadeath T-shirts, attire no other country would tolerate.

Moreover, American schools don't consider academic achievement their principal goal. Social development, emotional well-being, the instillation of democratic values, and athletics all take precedence over academics. Educators, from the Commissioner of Education to the classroom teacher, talk of the "children's needs," never the "students' responsibilities." In practice, this means that parents and schools don't demand the time and effort from their students that are necessary for academic excellence. One eighth-grade science teacher told me that her students spent the first half of the year raising money for the class trip, took a week off from school to visit Washington, D.C., worked on the school play during class time, were excused from classes to do community service (part of social studies, I presume), and finally took more class time to rehearse for graduation. The whole system seems designed to keep the students busy with anything but studying.

Most countries don't require school beyond ninth grade. This isn't because they're backward, but because a ninth-grade education is adequate for most people. If U.S. students knew 50 percent of what the eighth- and ninth-grade mathematics, science, history, and English textbooks contain, they too would be well educated. The problem is that most don't know anywhere near this much by ninth grade, by twelfth grade, or even after graduating from college.

By requiring attendance in school to age sixteen and measuring educational success by the percentage of students who graduate from high school, we have undermined the quality of public education. In addition, public education is drowning in a sea of mandated programs to mainstream students with a wide range of mental and physical problems. Special-needs students are mainstreamed into science classes with teachers untrained in either science education or special education. How does one challenge the best students while catering to the neediest? Is this possible even under ideal conditions, let alone in a real classroom?

Today, the poorest and least successful families dictate what public schools can demand of students and parents. A science teacher can't ask his students to have five-dollar calculators because the poorest may not be able to afford it. Homework is challenged by students whose chaotic home life provides neither space nor opportunity to study. The quality of education deteriorates as teachers stop challenging students who have stopped studying.

It may not be possible for every school to handle every child. If public schools can't demand minimum standards of behavior and ability, parents who can will send their children elsewhere. This will lower support for school budgets even more, making funding ever more problematic and the public schools ever more desperate. In the end, the special-needs legislation that requires schools to provide a full education for all students with mental, physical, or emotional problems—regardless of costs—will bankrupt public education, just as Medicare will bankrupt Social Security, since the needs of the neediest are infinite. At this point, the United States will have completed its transition to a competitive nation based on a low-wage working class, dominated by an upper class that sends its children to private or elite suburban public schools.

A Fantasy

In order to think about alternatives to the current educational system, let me fantasize a bit. Imagine an educational system that requires all students to complete a standard nine-year curriculum, culminating with algebra and geometry, physical science, English and American literature, and world history. After seventh grade, students enter an elite academy for two years of intensive study. Academy students are expected to devote themselves fully to their studies, with several hours of homework each night. School is continual for twenty-two months, with a four-week break in the middle.

Society recognizes the academy years as a very special time in young people's lives and supports their efforts by reducing other expectations. Students receive distinctive academy caps, and stores offer small discounts as a token of respect. The entertainment industry cooperates (remember, this is a fantasy) by providing positive role models for this age group. In other words, the academy period is treated with the seriousness of national service, and the students are constantly reminded that what they're doing is for themselves, their family, and their country.

The teachers, facilities, and equipment in the academy are first rate, with classes half the size of nonacademy classes. The academy doesn't make everyone a scholar, but it does provide an environment in which each student's potential is developed more fully than it is at present. And because the teaching staff is large, most academy students achieve the academy's minimum goals: to be handy with a calculator and a dictionary. Students who can't enter the academy because of emotional, mental, or physical problems are given an alternative education, at the same per capita funding level.

Graduation from the academy is the most important academic achievement in a person's life, signifying passage to adulthood. This is the heart of the fantasy. Adulthood is moved from some nebulous distant age—eighteen, twenty-one, thirty, . . . —to fifteen. This is the school leaving age in Britain and many other countries. There is no disgrace in it. On the contrary, graduation from the academy is respected because academy students work very hard. And they work very hard because they know adulthood is right around the corner. For this reason and because of the large teacher staff and society's wholehearted support, more than 90 percent of academy students graduate.

Further education is optional and is provided free to students who can benefit from it. Perhaps 90 percent of the academy graduates go on to a tenth year with continued academic programs for the college bound and office and vocational training for those who will be entering the work force. Cooperative education programs give some students on-the-job training in local industries.

In this fantasy, the public-school system is saved because it offers to most of its students a private school–quality education for a few years. And this fantasy gets even better, because it doesn't cost a penny extra. In fact, it costs less than the current public educational system. How, even in fantasy, is this done? Simple. By eliminating the last two years of high school.

Eliminate high school! Preposterous! Unthinkable! Outrageous! The destruction of the American way of life!

But is it? The average college student isn't comfortable with a calculator or a dictionary. Since seventh grade, they've had five very loose years of schooling in which they do everything but study. American public education is hugely inefficient. My fantasized academy is demanding and focused but tolerated by most students because it's for only two years. During these two years, students learn to concentrate and study, making easier whatever they do afterward. With a tenth year of further study, students would be better prepared for college than they are now.

A thousand objections come to mind. What happens to the high-school teachers? They, of course, are needed in the academy. Indeed, their expertise is vital to the quality of the academy's education. Aren't students too young to work so hard? The ages thirteen to fifteen are a time of great intellectual development and energy that is largely wasted in the United States. Students this age in other countries work much harder than Americans, thereby putting Americans permanently behind. Wouldn't the intensity of the academy destroy childhood? American society doesn't provide a well-defined transition to adulthood. Employers complain of workers who remain adolescent into their thirties. A well-balanced life doesn't required an extended childhood, but the effective integration of work and play. The academy provides the same quality transition that elite private schools do. It makes everyone privileged.

What about the football program? What about music, art, the school newspaper? There are a host of time-honored activities now supported by the taxpayer that would be eliminated in my fantasy. Owing to budget cuts, some have been already, without any improvement in ninth-grade education. More to the point, however, is the question of what activities the taxpayer should be forced to fund: football for the few who make the varsity squad or a high-quality learning experience for all?

In this fantasy, students start college at age sixteen and graduate at twenty. This may seem a radical departure from current practice. Aren't most sixteen-year-olds too immature for college? Wouldn't the social and academic pressure be too much for them? Certainly not all sixteen-year-olds are ready to live away from home, but most college students go to a local college or community college anyway. Since all students will have had intense academy training, they generally will be better prepared than today's students are.

The United States has had some experience with admitting bright students to college after two years of high school. In the 1950s, in an experiment sponsored by the Ford Foundation, 1,350 young men and women were admitted to twelve colleges after their sophomore and junior years of high school.[5] Although the experiment was originally conceived as a program to provide bright young men with two years of college before they were drafted into military service, it soon came to view itself as a large-scale experiment in early admission. The Ford Foundation's 1957 evaluation report pointed out:

> The experiment was one of a combination of five projects supported by the [foundation] as part of a broad-scale attack on two closely related weaknesses in the American educational system which tend to impair quality and impose waste. The first

is a lack of sufficient flexibility to accommodate the wide differences in ability, interests, and maturity that prevail among
young people of similar age. The second is a lack of continuity
in the various stages of the educational process, which too
often leaves gaps in a student's education or forces him to repeat work he has already done well. (Fund, 1957, p. 1)

I was among the first fifty Ford Scholars admitted to the Uni-
versit of Wisconsin in 1951, having completed just two years of
high school. Surprisingly, the only subject I lacked was trigonometry, and so I took a college-level trigonometry and analytic geometry course, which was excellent preparation for Calculus 1 in my
second semester (Cromer, 1989). Although I was a bright, hard-
working student, I came to Wisconsin from a deteriorating inner-
city high school that gave me no special boost up the academic lad-
der. In fact, except for trigonometry, my university courses expected me to know nothing from high school. All my introductory
science courses—biology, chemistry, and physics—started from
scratch, as did my literature and social studies courses.

Overall, the faculty and administrators at the twelve participating colleges judged that early admission was a "wise" choice for
more than 75 percent of the Scholars and "unwise" for less than 7
percent of them. The experiment proved that the last two years of
high school in themselves didn't enhance college performance. The
Scholars, as a group, did substantially better than a comparison group
of high-school graduates with matched SAT scores. Forty percent
of the early admission students were in the top fifth of their college
class, as opposed to 30 percent of their matched counterparts.

Today, many colleges have admission policies that can accommodate early-admission students on a case-by-case basis, though very
few have established a full-scale early-admission program or actively
recruit early-admission students. This ambivalent approach to early
admission was adopted, I believe, to avoid competing with secondary schools for the brightest students.

The wide availability of advanced placement courses in high
school is an acknowledgment that bright high-school students don't
need the last year or so of high school to prepare for college-level
work. But these courses only compound, at added cost, the "lack
of continuity" of the educational process that early admission re-
dresses. When they get to college, our best high-school graduates
find themselves either repeating a calculus course they had in high

school or competing with college-prepared sophomores in an advanced calculus course. They deserve better than this.

In 1988, the University of North Texas established the Texas Academy of Mathematics and Science, a two-year residential program for students who have completed two years of high school. Although these students occupy their own section of a university residence hall, they eat in the same cafeteria and take the same courses as the regular North Texas students do. Only their calculus course is special, having been modified to give them the trigonometry and analytic geometry they didn't study in high school.

Since 1988, the freshman class at the Texas Academy of Mathematics and Science has grown from 88 to over 200, and in 1993 the Texas Academy of Humanities opened at Lamar University. These academies are true early-admission programs. The students get college credit for college courses taught by college faculty. These programs are thus more progressive and much less expensive than the residential high schools of North Carolina and Louisiana. And the students are doing well, better than the average Texas freshmen. After two years, most academy students at North Texas transfer as juniors to other colleges and universities throughout Texas and the United States.

There is thus consistent evidence that eliminating the last two years of high school benefits the bright students. And there are many reasons that eliminating the last two years of high school and focusing these resources on the eighth and ninth grades would benefit everyone: First, all public-school students would receive at least two years of the highest-quality education. Second, schooling for everyone would be shortened by two years, giving everyone two extra years of employment at their highest salary. This is an increase of lifetime income of over 6 percent.[6] Third, taxpayers would spend somewhat less on public education. Fourth, parental support of children would be shortened by two years.

Although this is only a fantasy, the alternative is stark reality: a public-school system that collapses under a burden of impossible social mandates and reduced funding. In this scenario, the middle class struggles to get its children out of the public-school system to which it still owes taxes. Instead of reaping the windfall of two extra working years, the middle class gets socked with private-school tuition. And the devil takes the hindmost.

The truth is that the United States can afford anything it wants, but not everything it wants. Choices must be made. No meaningful progress toward school reform is possible without agreeing on a maximum agenda. The public schools cannot continue to be the venue for the teaching of every subject, the instilling of every public virtue, or the treating of every social ill. If they do, they will continue to lose support as concerned parents look to private schools more focused on academic excellence. Only by tapping the resources and talents of our high schools can we get out of the hole we are rapidly digging and provide a high-quality education for all.

Summary

Teaching science and mathematics requires relevant concrete activities designed for different levels of cognitive development.

Teachers should have opportunities to learn about these activities from one another in afterschool workshops.

Middle-school science teachers should have volunteers, preferably retired engineers and scientists, to help them prepare demonstrations and laboratories and to offer real-world examples to their students.

Education students should be taught science with the same activities that they will be using with their students.

Middle-school education should be enhanced by eliminating the eleventh and twelfth grades and transferring the teachers and facilities to the eighth and ninth grades.

Conclusion

Science is a new factor in human existence that goes against the grain of our egocentric mind. It originated in the democratic practices of ancient Greece, which replaced private dogma with public debate. Based on the heretical belief in an external world that can't be known intuitively, science replaced speculation and faith with reason and observation. Much of our progress, both political and technological, has come from accepting that truth can arise only from the competition of free and informed minds. It cannot be imposed by cannon or canon.

Over the centuries, science has flourished for a time in one culture or another, but in the end has always been eclipsed by ideology and superstition. Currently, science is supported by most nations because of its direct contribution to their technological and industrial advancement. However, those of us who value science shouldn't be complacent. Cults and prophets arise spontaneously from the depths of the egocentric mind, whereas scientific knowledge must be passed from generation to generation by a complex educational process that spans twenty years or more.

If the United States fails to provide a quality education for all, ignorance and superstition will grow rapidly. Already much public policy, from the regulation of low-frequency radiation to the content of science textbooks, is made by people with no understanding of science. Private schools and elite public schools will, for a time, be able to provide an excellent education to a privileged minority, but just as Europe once abandoned Greek education in a wave of pious fundamentalism, America may someday turn out the light.

The Disney movie *Beauty and the Beast* can be seen as a parable about science. This wonderfully animated fairy tale isn't set in some far-off time and place, but in late-seventeenth-century France. Mechanical clocks, based on Galileo's study of the pendulum, are available to the wealthy. Inexpensive printed books are sold even in Beauty's small village. And most important, her father is a sort of mad scientist, working on a fantastical steam-driven, wood-chopping contraption. Steam, in fact, became industrially important a few decades later, thanks to the work of seventeenth-century inventors like Beauty's father.

Beauty herself, bored with her provincial life, finds herself imprisoned in an enchanted castle where a sorceress has transformed the prince into a beast and all the servants into bric-a-brac. Although charmed by a singing candlestick and a matronly teapot, she eventually breaks the enchantment with her love for the Beast and some help from the steam machine.

As a lover of both fairy tales and science, I was disturbed by the idea, inherent in the movie, that science destroys enchantment. Science is a sort of growing up of humankind, replacing animistic fantasies with an objective external world, just as Beauty's love for the Beast changes her from a girl to a woman (Bettelheim, 1975).

But in the end, Beauty gets a real prince and a real castle, albeit no longer enchanted. Where is humankind's castle? The earth, once thought to be the center of existence, is now just a cinder in space. Has growing up deprived us of more than we gained?

I think not. With our current understanding, we realize that the earth is in fact a very special place. In a universe composed mostly of hydrogen and helium, the earth is rich in carbon and oxygen. Its thin shell of ocean and air contains the only life in our neighborhood of the galaxy, an enchanted bubble adrift in the dark emptiness of space.

The earth is our castle, animated with millions of singing birds and dancing butterflies, more marvelous than any Disney candlestick or teapot. Its storms and tempests fill us with awe, and its uniqueness and fragility fill us with reverence. Reality has far more wonders than all the tales of Arabia, giving us in return for our lost feeling of omnipotence some knowledge of the external world, some control over and responsibility for our lives, and even a touch of humility.

Appendix A

Hindu Trigonometry

The ratio a/h of the side a to the hypotenuse h of a right triangle is called the sine of θ, where θ is the angle of the triangle opposite the side a (see Figure 6.1). The sine varies from 0 to 1 as θ varies from 0° to 90°.

The ancients defined things a little differently. The Greeks arbitrarily fixed the value of the hypotenuse to be 60 and made tables of the corresponding values of a for different values of θ. For example, because the sine of 30° is 0.5, the Greeks would have $a(30°) = 30$, meaning that the side opposite the 30° angle is half the length of the hypotenuse.

The Greeks chose 60 for the hypotenuse because they used the Babylonian number system based on 60. For more precision, they would divide the length of the hypotenuse into $60 \times 60 = 3600$ divisions, called minutes. In this case, $a(30°) = 1800$ minutes. Because $\sin(45°) = 0.7071$, $a(45°)$ is $0.7071 \times 3600 = 2546$ minutes, approximately.

Ancient Hindu astronomical texts, the *Siddhantas,* used the Hindu decimal system, but divided the circle in the Greco-Babylonian way, into 360°. The *Surya Siddhanta* contains a table of sines for angles from 0 to 90° in steps of 3.75°. This is the angular interval to be expected for a table that was constructed by starting with $\sin(30°) = 0.5$, using elementary trigonometric identities to find the sine of successive half-angles (15°, 7.5°, 3.75°), and then using identities to find the sine of successive multiples of 3.75°. This, however, isn't the method that the *Surya Siddhanta* itself gives for calculating the table (Brennand, 1896).

The Hindu hypotenuse wasn't 3600, but 3438. This strange number has a valid mathematical derivation. It is the number of minutes in an arc of a circle equal to its radius, $(360 \times 60)/2\pi$. The use of this subdivision is related to the modern use of radians to measure angles. With it, $a(\theta) = \theta$ (approximately), for small θ. For example, $\theta = 3.75° = 3.75 \times 60 = 225$ minutes, and $a(225) = 225$ (approximately) if $h = 3438$. To compare with our sine, we divide a by $h = 3438$:

$$\sin (3.75°) = \frac{a(3.75°)}{h} = \frac{225}{3438} = 0.065445$$

This is to be compared with the exact value, $\sin(3.75°) = 0.065403$.

So far so good. The problem is with the formula the *Surya* gives for finding a for larger angles:

> The eighth part of the number of minutes contained in a sign [of the zodiac, i.e., 30°] is the first a. Divide the first a by itself, subtract the quotient from that a, and add the remainder to that a; the sum will be the second a. In the same manner, divide successively the a's found by the first a; subtract the (sum of) the quotients from the divisor, and the remainder to the last a found, and the sum will be the next a. Thus, you will get twenty-four a's (in the quadrant of a circle whose radius is 3438). (Brennand, 1896, p. 210)

All formulas in the *Siddhantas* are given in words, without proof or logical explanation. Although many are correct, this one isn't. In modern notation, it reads

$$a((n + 1)\theta) = a(n\theta) + a(\theta) - \frac{a(\theta) + a(2\theta) + \ldots + a(n\theta)}{a(\theta)}$$

To a mathematician this is a patently absurd equation. The a's are lengths, but the quotient of lengths is a pure number. The formula adds apples and pears. Consequently, it gives different results depending on the value chosen for the hypotenuse and for the first angle θ.

Reading again the description of the method in the *Surya*, we see that the author didn't view the initial angle as some arbitrary starting point, but as the eighth part of a sign—that is, a unique number of some mystical significance. In fact, that's how all the *Siddhantas* are written. They are filled with references to gods and demons and mystical cycles. Ancient rules mustn't be altered, but corrections could be made to fit current planetary positions.

Starting with $a(225) = 225$, the *Surya* formula yields $a(7.5°) = 449$, $a(10.25°) = 671$, and $a(15°) = 890$. By dividing each of these numbers by 3438, we get the sines. Thus $\text{sine}(15°) = 890/3438 = 0.2589$, compared with the exact value of 0.2588. Not bad. However by 90°, the formula yields $a(90°) = 3375$, or $\sin(90°) = 0.9817$, instead of 1.

It is remarkable that the formula does as well as it does. It doesn't work nearly as well for other values of the hypotenuse or starting angle. For example, instead of arbitrarily dividing the circle into 360° and $360 \times 60 = 21,600$ minutes, let us arbitrarily divide it into 100,000 parts. Then the hypotenuse h must be taken to be $100,000/2\pi = 15916$ and $\theta = 3.75°$ will be $(3.75/360) \times 100,000 = 1042$. With these values, the formula yields $a(30°) = 8248$, or $\sin(30°) = 8248/15916 = 0.518$ (compared with the exact value of 0.5), and $a(90°) = 22758$, or $\sin(90°) = 22758/15916 = 1.429$ (compared with the exact value of 1). Thus in general, the formula yields nothing like the correct values.

The conclusion of all of this is that the ancient Hindu astronomers had a correct table of sines, but no knowledge of how it was constructed. More important, the *Siddhantas* show no knowledge of mathematics as a system of logical deductions from basic premises. Rather, they treat mathematics, including the table of sines, as a system of divinely revealed formulas. It would appear that the formula in the *Surya* for the table of sines was concocted to fit a table that had been inherited from the Greeks centuries earlier, but without an understanding of its method of construction.

Appendix B

An Integrated Science Course for Nonscience Students

For two years (1991–1993), I taught Introduction to Science, a one-quarter science course that nonscience students could elect to satisfy their science requirement. The students in this course were from all classes (freshmen to senior) and all nonscience disciplines (business, criminal justice, journalism, liberal arts, and education). In an eleven-week quarter, the course told the story of cosmic and biological evolution, from the Big Bang to human civilization, and the story of how science arrived at its current understanding of these evolutions. Specifically, the course covered the scientific revolution, the laws of motion, stars and galaxies, elements and compounds, atomic and nuclear structure, cosmic evolution, stellar evolution, radioactivity, the structure of the earth, plate tectonics, organic chemistry, genetics, DNA, evolution, and human evolution. As preposterous as this sounds when summarized in this way, most of the material can be found in seventh-, eighth-, and ninth-grade physical science, earth science, and biology textbooks. I used a standard university-level physical science textbook (Krauskopf and Beiser, 1991), supplemented by notes (principally Chapter 3 of this book).

The course consisted of lectures, weekly homework assignments, five biweekly half-hour exams, and a two-hour final. Figure B.1 shows a typical final examination, tidied up a bit for publication. Although the test is technical in parts, similar questions were given on the biweekly exams. Each question was discussed before the biweekly exam in which it was first given, after that exam, and before the final. The class average on the final exam was usually about 68 percent (C−). This doesn't sound too bad until you remember that the course didn't go beyond what everyone should have learned by ninth grade.

What follows is a question-by-question analysis of the final examination in Figure B.1. The purpose is to show the cognitive level of average nonscience college students and how difficult it is to raise it in a lecture course. I prefer the short-response questions given here to either multiple-choice or essay questions, because it is possible to give partial credit to the responses with reasonable consistency.

1. The course starts with a discussion of the night sky and how the stars appear to move in circles about the North Star (Polaris) every twenty-four hours. To demonstrate comprehension, the students are asked to draw the path that two stars make in six hours. A typical mistake is drawing half-circles instead of quarter-circles; worse, a few students fail to recognize that star A, closer to Polaris than star B, must move in a smaller circle than star B does.

2. Scientific (power of ten) notation is necessary for dealing with very small and very large distances and times. Students have only minor troubles with the mechanics of this, but they don't know the number of zeros in a billion and a trillion. Few are able to answer this important question correctly, but because the test is hand graded, they get partial credit for the "correct approach."

3. Students have trouble with chronological order because they seldom relate one fact to another. Even though I give the students chronology charts and discuss the major figures, they are thrown by this simple history question when it is first given; Muhammad's role in history is a blank to all but my Arab students.

4. Students generally do well on this question about the earth's orbit.

5. An example of a time-rate-distance problem involving proportional thinking. I can get most of my students up to this level, but they fall apart when I ask them for the speed of a continent that moved 1,200 kilometers in 65 million years.

6. and 7. These conceptional questions about motion are more

Northeastern University
Introduction to Science
Final Examination

1. The following figure shows the position of Cepheus, the Big Dipper, the Little Dipper, and Polaris at 6:00 P.M. on a winter's night. On this figure, indicate the positions of stars A and B at midnight on the same evening and the path along which each star traveled between 6:00 P.M. and midnight.

2. a. Write the following in scientific notation:
150 thousand =
3.6 trillion =
250 million =

b. The national debt (1991) is $3.6 trillion, and the population of the United States is 250 million. Calculate the national debt *per capita*, that is, the debt "owed" by each individual (man, woman, and child) in the United States, assuming the debt were uniformly distributed.

3. List the following people in chronological order: **Copernicus, Darwin, Galileo, Lavoisier, Mohammed, Moses, Newton, Ptolemy.**

4. The following figure shows the earth at one time.
a. At what month of the year is the earth in this position.
b. Draw the position of the earth six months later. Be sure to show the position of the North Pole at this later time.

5. a. The earth, in its orbit around the sun, travels 9000 km in 5 minutes. What is the speed of the earth in km/hr?
b. How far does the earth travel in 24 hr?

6. The following figure shows a ball on the end of a string being twirled clockwise (from A to B) in a **horizontal** circle.
a. At point A, draw an arrow in the direction of the **acceleration** of the ball at that point.
b. At point B, the string breaks. From B, draw the path the ball following immediately after the string breaks. (Ignore gravity.)

7. A girl throws a ball stright up in the air. Neglecting air resistance, decribe
a. How the velocity of the ball changes while the ball is in the air.
b. How the acceleration of the ball changes while the ball is in the air.

8. Galileo was the first person to observe four of the moons of Jupiter. One of these moons, Europa, moves in a circular orbit of radius 0.67 million kilometer with a period of 3.56 days. Another, Ganymede, moves in a circular orbit of radius 1.07 million kilometers with a period of 7.18 days. Show that these two moons of the same planet obey Kepler's third law.

9. The following chart shows some of the properties of stars A and B. Fill in the blanks with the correct word (**larger, smaller,** or **same**).

Property	Star A	Star B
Apparent brightness	Same	Same
Distance		
Intrinsic brightness	Smaller	Larger
Color	Yellow	Red
Temperature		
Size		

10. Fill in the blanks in the following statements.
a. 9 g of carbon (C) combine with _____ g of hydrogen (H) to produce 12 g of methane (CH_4) and nothing left over.
b. _____ g of C combine with 5 g of H to produce _____ g of methane and nothing left over.
c. 6 g of C combine with 3 g of H to produce _____ g of methane and _____ g of _____.

11. Fill in the following chart with the appropriate words from the following list: **interstellar gas, H–He, H–He–C–Fe–etc., gravity, nuclear force, high temperature, supernova, white dwarf, red giant, fusion (p–p cycle).**

12. Fill in the following chart with the appropriate words from the following list: **atoms, atomic number, compound, electrons, element, mixture, law of definite proportions, molecules, neutron, nucleus, proton.**

Figure B.1 Final examination for Introduction to Science, a one–quarter course for nonscience students at Northeastern University.

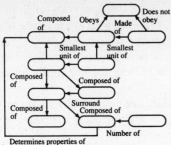

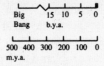

first occurred. Events that occurred within tha last 500 million years should be placed on the expanded time line (b.y.a. = billion years ago, and m.y.a. = million years ago).

```
 Big        15  10  5   0
 Bang      b.y.a.
```

```
 500  400  300  200  100  0
 m.y.a.
```

A. Creation of the sun
B. Extinction of the dinosaurs
C. First life on earth
D. First apes
E. Beginning of the Cambrian period
F. Creation of the first helium in the universe

Determines properties of

13. The actinides are:

Ac Th Pa U Np Pu Am Cm Bk ...
89 90 91 92 93 94 95 96 97 ...

a. Write the mass number and chemical symbol of the nucleus that results from two successive alpha decays of ^{237}Np.

b. Write the mass number and chemical symbol of the nucleus that results from two successive beta decays of ^{239}U.

14. ^{40}K decays to ^{40}Ar with a half-life of 1.3 billion years. The ^{40}Ar is stable. In the table below, give the number of ^{40}K and ^{40}Ar atoms after four successive periods of 1.3 billion years.

Time (billion years)	^{40}K (no. of atoms)	^{40}Ar (no. of atoms)
0	1000	0
1.3		
2.6		
3.9		
5.2		

15. Hexene (C_6H_{12}) is a hydrocarbon with one double bond. Since this double bond may be in several different places, there are several isomers of hexene. Draw the structural formulas for two isomers of hexene.

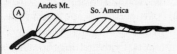

Andes Mt.   So. America
(A)

16. On the above diagram, identify by letter each of the items listed below. The first item is done for you as an example.
A. Pacific Ocean
B. A ridge where material is being added to the earth's surface
C. A subduction zone
D. The South American plate (indicate its eastern and western limits with arrows labeled D)
E. The direction of motion of the South American plate (indicated with an arrow labeled E)

17. By letter, indicate on the following time lines approximately when each of the events listed below

18. The sequence of bases –CTAGCTTTCATT ... etc. –on one strand of DNA codes for the sequence of amino acids leucine, alanine, phenylalanine ... etc.

a. What is the genetic code for leucine, alanine, and phenylalanine?

b. What is the corresponding sequence of bases on the matching strand of DNA?

19. Fill in the following chart with the appropriate words from the following list: **amino acid, cell, chromosome, evolution, gene, inheritance, mutation, natural selection, protein, variation.**

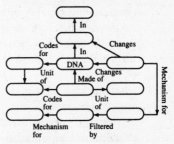

20. By letter, indicate on the following time lines approximately when each of the events listed below first occurred. Events that occurred within the last 200,000 years should be placed on the expanded time line.

```
 5    4    3    2    1    0
 m.y.a.
```

```
 200,000      100,000        0
 y.a.
```

A. Humans replaced Neanderthals
B. Fire
C. Agriculture
D. Bipedalism
E. Stone tools
F. *Homo sapiens*

difficult than they appear because students come to the subject with many misconceptions (Hestenes, Wells, and Swackhamer, 1992). Student performance was expectedly low, given the small amount of time devoted to the subject in the course.

8. This problem on Kepler's third law is easier than it appears, since it involves a straightforward procedure that most students eventually memorize.

9. This question tests the students' understanding of the physics and logic involved in determining the size of stars. They tend to think that red giants (large red stars) are large because they're giants. Even if students don't fully understand what this question is about, they begin to distinguish observations from inferences.

10. I stress the law of definite proportions because of its great historical importance and the opportunity it offers to ask questions about proportions. According to Piaget, proportional thinking is a formal operation not accessible to students at the concrete operations stage. All my students can do the first line, which involves only subtraction. Half learn to do the second, which is a simple proportion. But few can do the third, which involves both a proportion and a subtraction.

11. I got the idea of concept maps from a ninth-grade physical science textbook (Carle, Sarquis, and Nolan, 1991), which teaches students to make their own maps. Such maps are excellent for summarizing the relationships among a large number of terms. I used this particular concept map in class to summarize cosmic and stellar evolution. A few ovals are filled in to keep the students from getting hopelessly lost.

12. A concept map on atomic theory. Students do well on this.

13. and 14. Radioactivity is covered in the course because of its use in geological and archaeological dating. Some students have a hard time with Problem 13 because they don't grasp atomic and nuclear structure. Although at one level this structure is very concrete, students have a hard time thinking about structures one-trillionth the size of the head of a pin. They do better with Problem 14.

15. Students do well drawing hydrocarbons. They have no

trouble drawing one isomer, but often draw the second one as a bent version of the first.

16. This is a straight-forward identification question based on a picture in the textbook. Students do surprisingly poorly on it, probably because they try to memorize terms detached from their meaning. (I don't know much about subduction zones myself, but I know one when I see one.)

17. and 20. Chronological charts are given to the students at the beginning of the course and are frequently discussed. Students are confused by the different time scales involved.

18. This question tests a little about the structure of DNA and the nature of the genetic code. Most students get this right.

19. This concept map encapsulates a lot of ideas from the end of the course. It is meant to teach the students as much as test them, but I am not sure it's successful at either.

This course was intended to be a unified summary of what the students should have learned in the seventh, eighth, and ninth grades. But since my students never learned middle-school science, my attempt at synthesis was largely wasted on them. What they need is a solid ninth-grade science course with a laboratory that provides concrete experiences with acquiring and processing quantitative information.

The last time I taught Introduction to Science (spring 1993), I replaced six of my lectures (out of twenty-seven) with six one-hour laboratory sessions. This required my eliminating about 30 percent of the topics shown in Figure B.1. In the laboratory sessions, the students worked in teams of three and handed in a one-page team report at the end of each session. This protocol was very effective in getting the students to engage fully with the material and with one another. They needed suprisingly little supervision; with the assistance of a graduate student, I easily ran laboratory sessions with fifty students.

The laboratory sessions involved simple activities such as measuring the periods of pendulums of different lengths and the average speeds at which one, two, four, eight, and ten coffee filters fall. They were, to the best of my ability, solid ninth-grade activities. Although many of the sudents had studied mathematics through calculus, they had had so little practice with graphing, fractions, and simple proportions that these were a struggle for them. However, by working together in groups, the students overcame these obstacles, which resulted in good work and a successful learning experience—at a ninth-grade level.

I conclude from this that little cognitive development is taking place in high school or college, and that science "reform" efforts aimed at teaching broad themes rather than specific skills will only perpetuate our current "content-free" educational system.

Notes

Chapter 1

1. Henry Cavendish was the actual discoverer of the composition of water. Lavoisier is credited with establishing the general principles of combustion from his own experiments and those of others (often without attribution). He died on the guillotine in 1794, during the Reign of Terror (Jaffe, 1930).

2. Because of their importance, the isotopes of hydrogen are given special names. Hydrogen-2 is called deuterium, and the hydrogen-3 is called tritium.

3. It is estimated that there is only an ounce (30 g) of astatine in the entire earth's crust.

4. German researchers have proposed the names nielsbohrium for element 107, hassium (Latin for Hesse, the German city in which the element was produced) for element 108, and meitnerium (after Lise Meitner) for element 109.

5. The bases in DNA are adenine, thymine, guanine, and cytosine.

Chapter 3

1. In 1992, the National Zoo in Washington, D.C., had a one-year-old gorilla. Although it is difficult to see much resemblance between ourselves and a big, old silverback male gorilla, the one-year-old gorilla was astonishingly similar to a human child of the same age. This supports the theory that human beings are juvenile apes—that is, that *H. sapiens* is an ape that retains juvenile characteristics into adulthood (neoteny) (Gould, 1979, pp. 63–69).

2. The "out of Africa" scenario given here isn't accepted by all anthropologists (Templeton, 1993). A competing scenario, called regional conti-

nuity, has modern humans evolving independently in different parts of Africa and Eurasia and forming a single species through interbreeding. But I am betting against regional continuity because it is contrary to my understanding of evolution as a divergent, not a convergent, process.

Chapter 4

1. Akhenaten is perhaps the first individual known to history. His life story is revealed in the inscriptions inside his tomb and the tombs of his family and ministers.

2. Michael Ventris was killed in an automobile accident in 1956, the same year his collaborative work with John Chadwick, *Documents in Mycenaean Greek,* was published.

3. Interestingly, some of the contradictory verses in the story of David and Goliath were omitted from the Greek translation of the Old Testament. Presumably the Hellenized Jews of Alexandria had become sensitized to matters of literary consistency (New Jerusalem Bible, 1985, p. 379, note k.)

4. In the *Iliad,* the right of free speech isn't unlimited, however, for Thersites does receive a blow from Odysseus for his insolence to the king.

5. In his ideal state, Plato would have passages such as this one eliminated from the version of Homer taught to young men (*Republic,* bk. iii; in Buchanan, 1948, p. 371).

Chapter 5

1. The discovery of irrational numbers held up the advancement of Greek mathematics until Eudoxus (fl. 365 B.C.) developed a new definition of equal ratios that applied equally to rational and irrational quantities. This definition is essentially the same as that used by Weierstrass for equal numbers and by Dedekind in the modern theory of irrationals. It comes down to us through Euclid (fl. 300 B.C.) as definition 5 in Book v of his famous work, the *Elements* (Heath, 1908, p. 124). See also Chapter 7, note 1.

2. This is an example of Plato's use of rhetorical exaggeration (*Laws;* quoted in Heath, 1921, vol. 1, p. 156).

3. To prove that $\sqrt{2}$ is irrational, we start by assuming that it is rational. That is, we assume that it can be expressed as a ratio n/m, where n and m are whole numbers with no common factors. In particular, n and m aren't both even, since two even numbers have 2 as a common factor.

But if $\sqrt{2} = n/m$, then $2 = n^2/m^2$ or $n^2 = 2m^2$ From this we see that n^2, and hence n, must also be even. But if n is even, the same equation shows that m is also even. Thus the assumption that $\sqrt{2}$ can be written as a ratio of two whole numbers is self-contradictory, and so $\sqrt{2}$ is irrational.

4. *Almagest* is the Latin form of the Arabic name Al-majisti, which is a combination of the Arabic *Al* (The) and the Greek *magistos* (greatest) (Heath, 1921, vol. 2, pp. 273–74).

Chapter 6

1. As a teacher and textbook writer, I have been particularly impressed by how rapidly the calculator replaced sine tables that had been in use for over 2,000 years. In *Physics for the Life Sciences* (Cromer, 1977), students were expected to work trig problems using a sine table printed in the back of the book, but in *Physics in Science and Industry* (Cromer, 1980) there was no longer a need for such a table. By 1980, all students had, and were expected to use, scientific calculators that give the sine or cosine of any angle at the push of a button.

2. In modern notation, the relation is

$$\sin\frac{\theta}{2} = \sqrt{\frac{1-\cos\theta}{2}} = \frac{\sqrt{1-\sqrt{1-\sin^2\theta}}}{\sqrt{2}}$$

With $\sin(30°) = 0.5$, this gives $\sin(15°) = 0.258819$.

3. The Greeks actually worked with chords, not sines. A chord is equal to twice the sine of the half-angle. Where I have written [sines] in the quotation, Ptolemy wrote "lengths of the chords."

4. In modern notation, this says

$$\sin \phi = s/h$$

where

$$h = \sqrt{(R \pm c)^2 + s^2}, \; s = r\sin\theta, \text{ and } c = r\cos\theta$$

5. Brennand (1896) incorrectly believed that the Hindu rule is a valid trigonometric relationship. He checked the rule for the first five angles, for which it does agree with the sine to four decimal places. But as shown in Appendix A, the rule gives larger errors for larger angles.

6. To call this device a seismograph, as Needham (1969) does, makes it sound more like a scientific instrument than it was. Likewise, it is misleading to call someone an astronomer—no matter how carefully he observes the stars—if his sole purpose is fortune-telling.

7. On their second voyage to India, in 1500, the Portuguese deliberately sailed southwest and discovered Brazil. Thus with their systematic program of exploration, the Portuguese should have been the first Europeans to reach America, but for the reckless and misguided Columbus, who beat them by eight years.

Chapter 7

1. A proportion is the relation among four quantities of the form "A is to B as C is to D." Every quantitative statement is a proportion, because even the simple statement "This steak weighs 3.5 pounds" means "The weight of this steak is to the weight of the 1-pound standard weight in

Washington as the number 3.5 is to the number 1." The problem is to come up with a definition of proportion that includes cases in which the steak weighs $\sqrt{2}$ pounds. The famous Definition 5 of Eudoxus, as given in Book v of Euclid, is: A is to B as C is to D if, for any two whole numbers n and m, $nA > mB$ implies $nC > mD$; $nA = mB$ implies $nC = mD$; and $nA < mB$ implies $nC < mD$.

2. The force of the table is not the reaction to gravity; it is an entirely independent force. Gravity does have a reaction force, as does the force of the table. Thus the simplest problem of a block at rest on a table involves four forces. I have never known a Ph.D. candidate, including myself, who, after five years of graduate study in physics, could correctly name the two forces and their reactions acting on a block at rest on a table.

3. The shifts in the positions of the nearest stars as the earth orbits the sun are 100 times smaller than what can be seen with the unaided eye. The first stellar shift was observed in 1838, when sufficiently powerful telescopes were available. The distance to the nearest star is more than 100,000 times the diameter of the earth's orbit.

4. Simplicio is named after Simplicius, a sixth-century Aristotelian commentator; the name does not connote simple-mindedness.

5. The earth's rotation does affect the motion of moving air, controlling such things as the circulation of hurricanes and the global flow of air from west to east in the middle latitudes of the Northern and Southern Hemispheres. The term *Coriolis* refers to the effect that the earth's rotation has on moving bodies.

Chapter 8

1. N people can be paired in $1/2N(N-1)$ ways. The factor of one-half ensures that the Jones–Smith interaction and the Smith–Jones interaction are counted as one. With $N = 100$, the number of pairs is 4,950, and with $N = 1000$, it is 499,500.

2. What Sagan here calls "borderline science" is better termed "pseudoscience" or "pathological science." In current terminology, the phrase "borderline science" is used for bona fide research into areas that most scientists think are unproductive. The search for extraterrestrial intelligence, in which Sagan takes great interest, is an example of borderline or fringe science (see Chapter 9).

3. Stuart's 6 hits per 25 is only 4 hits per 100 above chance. One hundred is 20 sets of 5. In every 20 sets, 6 on the average will have 3 or more cards of the same kind. Suppose that for these "unusual" sets the cards were shuffled and cut so that many of the same cards were again on top of the deck—an easy thing to do either consciously or unconsciously in a deck of 25 cards. Then by merely guessing the long suit of the previous hand, the extra 4 hits could be easily obtained. Although there is no way of knowing what Stuart did, the fact that it is easy to find plausible explanations for his results discredits the work.

4. At first, the "different" particles were called τ (tau) and θ (theta).

Now the single parity-violating particle is called the K meson. It comes in charged and uncharged varieties and is a wonderfully arcane beast that lives, from its creation to its destruction, for about ten-billionths of a second.

5. The name of the third author, M. Hawkins, was omitted on the original paper but was included in the erratum published later (Fleischman, Pons, and Hawkins, 1989).

6. Steven Jones (1992) is continuing his study of anomalous neutrons from deuterium in solids, although he is mindful that lack of repeatability and the very few neutrons involved is "a red flag that we may be chasing noise."

7. On January 2, 1992, one scientist was killed and three were injured when the deuterium gas used for cold-fusion experiments exploded at the SRI research facility in Menlo Park, California (Dye, 1992). Deuterium gas, like ordinary hydrogen, reacts explosively with oxygen to produce water.

8. This quotation is the gist of a conversation between Peter Scheubelin and Bernard Gottschalk, as recollected by Gottschalk. At the time, Scheubelin was defending the CERN results and begging Gottschalk, for Gottschalk's own good, not to present the Northeastern results.

Chapter 9

1. The 1,000-foot radio telescope at the Arecibo Observatory in Puerto Rico can transmit brief radio signals that could be detected by a similar telescope 15,000 light-years away (Kerr, 1992). However, the dish of the telescope is built into a bowl-shaped valley and always points straight up, a direction in space that changes continuously as the earth rotates on its axis and orbits around the sun. It could detect a signal from a similar telescope on another planet only during the brief and unlikely moment the two were exactly aligned.

2. The strength of a radio signal received on earth decreases as the square of the distance of the source from earth. Thus a source 1,000 times farther from earth than *Voyager 1* would have to be 1,000 squared (1,000 × 1,000), or 1 million, times stronger to deliver the same signal, all other factors being equal. A TAEC could do somewhat better using a larger transmitting antenna to produce a more directed signal, but only if it knew in which direction to aim it.

3. By 1892, the New York Central's famous engine 999 had reached speeds of 112.5 miles per hour while pulling the Empire State Express (White, 1972). Today, speeds in excess of 100 miles per hour are attained on the high-speed trains of France and Japan and on the Autobahn of Germany.

4. The Centauri Proxima triplet of stars are the stars closest to earth, but stable planetary orbits probably can't exist in the complex gravitational field produced by the trio.

5. It is shown in first-year college physics (Halliday and Resnick, 1988, p. 193) that the final speed of a rocket is equal to the speed at which the propellant leaves the rocket (exhaust speed, $V_{exhaust}$) times the natural logarithm of the mass of the fuel divided by the mass of the payload, or

$$V_{\text{rocket}} = V_{\text{exhaust}} \ln \frac{m_{\text{fuel}}}{m_{\text{payload}}}$$

Even with the mass of the fuel 1,000,000 times the mass of the spacecraft (payload), the final speed is only $\ln(1,000,000) = 13.8$ times the exhaust speed. The exhaust speed of the best liquid fuel (hydrogen–oxygen) is less than 3 miles per second (Ritchey, 1977), so even with a fuel-to-payload ratio of 1,000,000:1, the final rocket speed would be less than 41 miles per second.

Chapter 10

1. For example, half the members of a jury might not see the fallacy in the argument "The murderer had to have been very strong. The defendant is very strong. Therefore, the defendant is the murderer."

2. A tenth-grade biology teacher told me that after she gave a sex-education lecture, a student asked her what the word *abstinence* meant. Thus it's not only in science that we fail to communicate by using overly abstract language.

3. The National Research Council of the National Academy of Sciences has also gotten into the science-education business, hiring a committee of consultants with money from various federal agencies to write the *National Science Education Standards* (National Research Council, 1993). Like Project 2061, the aims of this committee are diffuse, diverse, and pretentious. Commenting on a first draft of its report, Clifford Swartz (1993), the distinguished editor of the *Physics Teacher*, wrote: "It is not a matter of changing a sentence here or there. This material has all the characteristics of curricula written by people specializing in general science education. The whole tone is wrong. The writers—or the compilers—seem to have no understanding of appropriate age levels for concept development, and there are enough infelicities in physics statements to make us suspicious that they don't understand the science either."

4. The principal investigators are Michael B. Silevitch, Christos Zahopoulos, and Alan Cromer. The project is supported by Northeastern University and a grant from the National Science Foundation.

5. In 1951, 348 young men and 72 young women were admitted to eleven colleges and universities (Chicago, Columbia, Fisk, Goucher, Lafayette, Louisville, Oberlin, Shimer, Utah, Wisconsin, and Yale). Morehouse joined in 1952.

6. Compare Ms. White, who starts her professional career at age twenty-three, with Mr. Black, who starts his at age twenty-five. Each receives a starting salary of $35,000 a year and a 1 percent real-wage increase each year. When Ms. White reaches age sixty-five, she will have earned $106,000 (6 percent) more than Mr. Black will have earned when he reaches age sixty-five. If they save 5 percent of their salaries at 3 percent real interest, when Ms. White reaches sixty-five, she will have $15,000 (9 percent) more in savings than Mr. Black will have when he reaches sixty-five.

References

Chapter 1

Armbruster, Peter, and Gottfried Münzenberg. 1989. "Creating Super-heavy Elements." *Scientific American,* May, 66.

Bloom, Allan. 1987. *The Closing of the American Mind.* New York: Simon and Schuster.

Clarke, I. F. 1979. *The Pattern of Expectation.* New York: Basic Books.

Hurd, Dean, Myrna Silver, Angela Bornn Bacher, and Charles William McLaughlin. 1991. *Physical Science.* Englewood Cliffs, N.J.: Prentice-Hall.

Jaffe, Bernard. 1930. *The Great Chemists.* New York: Tudor.

Kuhn, Thomas S. 1970. *The Structure of Scientific Revolutions.* 2nd ed., enlarged. Chicago: University of Chicago Press.

Mack, John. 1992. "Other Realities: The 'Alien-Abduction' Phenomenon." *Noetic Sciences Review,* Autumn, 5.

Rafelski, J., and S. E. Jones. 1987. "Cold Nuclear Fusion: The Electronlike Particles Called Muons Can Catalyze Nuclear Fusion Reactions." *Scientific American,* July, 84.

Sarton, George. 1960. *A History of Science.* Vol. 1. Cambridge, Mass.: Harvard University Press.

Velikovsky, Immanuel. 1950. *Worlds in Collision.* New York: Macmillan.

Watson, J. D., and F. M. C. Crick. 1953a. "Genetical Implications of the Structure of Deoxyribonuclei Acid." *Nature* 171:964.

Watson, J. D., and F. M. C. Crick. 1953b. "Molecular Structure of Nucleic Acid." *Nature* 171:737.

Ziman, John. 1968. *Public Knowledge.* Cambridge: Cambridge University Press.

Chapter 2

Benedict, Ruth. 1934/1989. *Patterns of Culture*. Boston: Houghton Mifflin.

Chandler, David. 1992. "UFO Reports Get a Going Over." *Boston Globe*, 22 June, 27.

Cromer, Alan, Christos Zahopoulos, and Michael Silevitch. 1992. "Chaos in the Corridor." *Physics Teacher* 30:382.

Galilei, Galileo. 1632/1967. *Dialogue Concerning the Two Chief World Systems—Ptolemaic and Copernican*. 2nd rev. ed. Ed. and trans. Stillman Drake. Berkeley: University of California Press.

Gardner, Martin. 1988. "Reich the Rainmaker: The Orgone Obsession." *Skeptical Inquirer* 13:26.

Gould, Stephen Jay. 1989. *Wonderful Life: The Burgess Shale and the Nature of History*. New York: Norton.

Gray, Thomas. 1987. "Educational Experience and Belief in Paranormal Phenomena." In *Cult Archaeology and Creationism*, ed. Francis B. Harrold and Raymond A. Eve. Iowa City: University of Iowa Press.

Hsu, Francis L. K. 1983. *Exorcising the Trouble Makers*. Westport, Conn.: Greenwood Press.

Lawson, Anton E., and John W. Renner. 1974. "A Quantitative Analysis of Responses to Piagetian Tasks and Its Implications for Curriculum." *Science Education* 58:545.

Piaget, Jean. 1929/1960. *The Child's Conception of the World*. London: Routledge & Kegan Paul.

Plummer, Mark. 1988. "CSICOP Challenges Mrs. Reagan's Astrologer to Scientific Test." *Skeptical Inquirer* 13:9.

Saltus, Richard. 1989. "U.S. Students Unprepared for College Science." *Boston Globe*. 26 February, 1.

Shinbrot, Troy, Celso Grebogi, Jack Wisdom, and James A. Yorke. 1992. "Chaos in a Double Pendulum." *American Journal of Physics* 60:491.

Wadsworth, Barry J. 1971. *Piaget's Theory of Cognitive Development*. New York: McKay.

Chapter 3

Darwin, Charles. 1859/1979. *On the Origin of Species by Means of Natural Selection, or the Preservation of Favoured Races in the Struggle for Life*. New York: Avenel Books.

de Waal, Frans. 1989. *Chimpanzee Politics*. Baltimore: Johns Hopkins University Press.

Diamond, J. 1987. "The Worst Mistake in the History of the Human Race." *Discover*, May, 64.

Dillehay, Tom D. 1991. "The Great Debate on the First Americans." *Anthropology Today*, August, 12.

Foley, R. A., and P. C. Lee. 1989. "Finite Social Space, Evolutionary Pathways, and Reconstructing Hominid Behavior." *Science* 243:901.

Gibbons, Ann. 1992a. "Human Ancestors Found—In Museum." *Science* 255:1071.

Gibbons, Ann. 1992b. "Mitochondrial Eve: Wounded, but Not Dead Yet." *Science* 257:873.

Goldstein, Melvyn, and Cynthia Beall. 1989. "The Remote World of Tibet's Nomads." *National Geographic,* June, 752.

Goodall, Jane. 1986. *The Chimpanzees of Gombe.* Cambridge, Mass.: Harvard University Press.

Gould, Stephen Jay. 1979. *Ever Since Darwin.* New York: Norton.

Gould, Stephen Jay. 1980. *The Panda's Thumb.* New York: Norton.

Gould, Stephen Jay. 1989. *Wonderful Life: The Burgess Shale and the Nature of History.* New York: Norton.

Irvine, William. 1959. *Apes, Angels, and Victorians.* New York: Meridian Books.

Johanson, Donald, and Maitland Edey. 1981. *Lucy: The Beginnings of Humankind.* New York: Simon and Schuster.

Josephus. 90/1970. *The Jewish War.* Rev. ed. Trans. G. A. Williamson. New York: Dorset Press.

Kehoe, Alice B. 1987. "Scientific Creationism: World View, Not Science." In *Cult Archaeology and Creationism,* ed. Francis B. Harrold and Raymond A. Eve. Iowa City: University of Iowa Press.

Kerr, Richard A. 1992. "Huge Impact Tied to Mass Extinction." *Science* 257:878.

Köhler, Wolfgang. 1927/1956. *The Mentality of Apes.* 2nd ed. London: Routledge & Kegan Paul.

Lewin, Roger. 1984. *Human Evolution.* New York: Freeman.

Lewin, Roger, 1987a. "Africa: Cradle of Modern Humans." *Science* 237:1292.

Lewin, Roger, 1987b. "Four Legs Bad, Two Legs Good." *Science* 235:969.

Lewin, Roger, 1988. "Modern Human Origins Under Close Scrutiny." *Science* 239:1240.

Macdonald, David, ed. 1984. *Primates.* New York: Torstar Books.

Morell, Virginia. 1987. "Announcing the Birth of a Heresy." *Discover,* March, 26.

Small, Meredith F. 1990. "Political Animal." *The Sciences* 30:36.

Stone, Irving. 1980. *The Origin.* New York: Doubleday.

Stringer, C. B., and P. Andrews. 1988. "Genetic and Fossil Evidence for the Origin of Modern Humans." *Science* 239:1263.

Swisher, C. C., et al. 1992. "Coeval [40]Ar/[39]Ar Ages of 65.0 Million Years Ago from Chicxulub Crater Melt Rock and Cretaceous–Tertiary Boundary Tekites." *Science* 257:954.

Templeton, Alan. 1993. "The 'Eve' Hypotheses: A Genetic Critique and Reanalysis." *American Anthropologist* 95:51.

Valladas, Héléne, Georges Valladas, Ofer Bar-Yosef, and Bernard Vandermeesch. 1991. "Thermoluminessence Dating of Neanderthal and Early Modern Humans in the Near East." *Endeavour* 11:115.

Waldrop, M. Mitchell. 1987. "The Workings of Working Memory." *Science* 237:1564.

Chapter 4

Bowen, James. 1972. *A History of Western Education*. Vol. 1: *The Ancient World*. New York: St. Martin's Press.

Buchanan, Scott, ed. 1948. *The Portable Plato*. Harmondsworth: Penguin Books.

Chadwick, John. 1968. *The Decipherment of Linear B*. 2nd ed. Cambridge: Cambridge University Press.

Freud, Sigmund. 1939/1955. *Moses and Monotheism*. New York: Random House.

Homer. 1946. *The Odyssey*. Trans. E. V. Rieu. Harmondsworth: Penguin Books.

Homer. 1950. *The Iliad*. Trans. E. V. Rieu. Harmondsworth: Penguin Books.

Koran. 1974. Trans. N. J. Dawood. Harmondsworth: Penguin Books.

The New Jerusalem Bible. 1985. New York: Doubleday.

Nilsson, Martin P. 1933. *Homer and Mycenae*. London: Methuen.

Weigall, Arthur E. P. 1910. *The Life and Times of Akhnaton*. London: William Blackwood.

Wood, Michael. 1985. *In Search of the Trojan War*. London: British Broadcasting Corporation.

Chapter 5

Aaboe, Asger. 1964. *Episodes from the Early History of Mathematics*. Washington, D.C.: Mathematical Association of America.

Berggren, J. L. 1986. *Episodes in the Mathematics of Medieval Islam*. New York: Springer-Verlag.

Boorstin, Daniel. 1985. *The Discoverers*. New York: Random House.

Chase, Arnold Biffum, H. P. Manning, and R. C. Archibald. 1927. *The Rhind Mathematical Papyrus*. Vol. 1: *Free Translation and Commentary*. Oberlin, Ohio: Mathematical Society of America.

Heath, Thomas, trans. 1908. *The Thirteen Books of Euclid's Elements*. Vol. 2. Cambridge: Cambridge University Press.

Heath, Sir Thomas. 1921. *A History of Greek Mathematics*. Vols. 1 and 2. Oxford: Oxford University Press.

Hoyle, Fred. 1975. *Astronomy and Cosmology*. New York: Freeman.

Hoyt, Edwin P. 1975. *Arab Science*. Nashville: Thomas Nelson.

Jaynes, Julian. 1976. *The Origin of Consciousness in the Breakdown of the Bicameral Mind*. Boston: Houghton Mifflin.

Kaufmann, William J., III. 1991. *Universe*. 3rd ed. New York: Freeman.

Nakayama, Shigeru. 1969. *A History of Japanese Astronomy*. Cambridge, Mass.: Harvard University Press.

Neugebauer, Otto. 1957. *The Exact Sciences in Antiquity*. Providence, R.I.: Brown University Press.

Pritchard, James B. 1958. *The Ancient Near East*. Vols. 1 and 2. Princeton, N.J.: Princeton University Press.

Sarton, George. 1960. *A History of Science*. Vol. 1. Cambridge, Mass.: Harvard University Press.

Chapter 6

The Bhagavad Gita. 1972. Trans. Franklin Edgerton. Cambridge, Mass.: Harvard University Press.

Boorstin, Daniel. 1985. *The Discoverers*. New York: Random House.

Brennand, W. 1896. *Hindu Astronomy*. London: Chas. Straker & Sons.

Chai, Ch'u, and Winberg Chai, eds. and trans. 1965. *The Sacred Books of Confucius and Other Confucian Classics*. New Hyde Park, N.Y.: University Books.

Coates, Austin. 1972. *China, India and the Ruins of Washington*. New York: Day.

Coomaraswamy, Ananda K., and Sister Nivedita. 1913/1967. *Myths of the Hindus and Buddhists*. New York: Dover.

Cromer, Alan. 1977. *Physics for the Life Sciences*. 2nd ed. New York: McGraw-Hill.

Cromer, Alan. 1980. *Physics in Science and Industry*. New York: McGraw-Hill.

Giles, Herbert A. 1901. *A History of Chinese Literature*. London: Heinemann.

Heath, Sir Thomas. 1921. *A History of Greek Mathematics*. Vol. 2. Oxford: Oxford University Press.

Mason, Stephen F. 1962. *A History of the Sciences*. New York: Macmillan.

Nakayama, Shigeru. 1969. *A History of Japanese Astronomy*. Cambridge, Mass.: Harvard University Press.

Needham, Joseph. 1969. *The Grand Titration*. Toronto: University of Toronto Press.

The Ramayana & The Mahabharata. 1910/1969. Trans. Romesh C. Dutt. Reprint. London: Everyman's Library.

Singer, Charles, et al., eds. 1956. *A History of Technology*. Vol. 2. Oxford: Oxford University Press.

Chapter 7

Boorstin, Daniel. 1985. *The Discoverers*. New York: Random House.

Bowen, James. 1972. *A History of Western Education*. Vol. 1: *The Ancient World*. New York: St. Martin's Press.

Butterfield, Herbert, 1965. *The Origin of Modern Science, 1300–1500*. New ed. New York: Macmillan.

Cromer, Alan. 1980. *Physics in Science and Industry*. New York: McGraw-Hill.

Ferris, Timothy. 1988. *Coming of Age in the Milky Way*. New York: Doubleday.

Galilei, Galileo. 1632/1967. *Dialogue Concerning the Two Chief World Systems—Ptolemaic and Copernican*. 2nd rev. ed. Ed. and trans. Stillman Drake. Berkeley: University of California Press.

Hawking, Stephen. 1988. *A Brief History of Time*. New York: Bantam.

Jeans, James. 1951. *The Growth of Physical Science*. 2nd ed. Cambridge: Cambridge University Press.

Heath, Thomas, trans. 1908. *The Thirteen Books of Euclid's Elements*. Vol. 2. Cambridge: Cambridge University Press.

Heath, Thomas. 1913. *Aristarchus of Samos*. Oxford: Oxford University Press.

Landau, Edmund. 1928/1951. *Foundations of Analysis*. Trans. F. Steinhardt. N.p.: Chelsea.

Needham, Joseph. 1969. *The Grand Titration*. Toronto: University of Toronto Press.

Newton, Isaac. 1687/1947. *Mathematical Principles of Natural Philosophy*. 1729 translation by Andrew Mott, revised by Florian Cajori. Berkeley: University of California Press.

Piaget, Jean, and Rolando Garcia. 1989. *Psycogenesis and the History of Science*. Trans. Helga Feider. New York: Columbia University Press.

Singer, Charles, et al., eds. 1956. *A History of Technology*. Vol. 2. Oxford: Oxford University Press.

Strayer, Joseph R., and Dana Carleton Munro. 1942. *The Middle Ages, 395–1500*. New York: Appleton-Century-Crofts.

Tuchman, Barbara W. 1978. *A Distant Mirror*. New York: Ballantine.

Chapter 8

Blackburn, T. R. 1971. "Sensuous–Intellectual Complementarity in Science." *Science* 172:1003.

Blackmore, Susan. 1987. "The Elusive Open Mind: Ten Years of Negative Research in Parapsychology." *Skeptical Inquirer* 11:244.

Bowen, D., et al. 1971. "Measurement of the A_2- and A_2+ Mass Spectra." *Physical Review Letters* 26:1663.

Chandler, David L. 1989a. "Optimism Grows over Fusion Report." *Boston Globe,* 1 April, 1.

Chandler, David L. 1989b. "Reports Set Off Chain Reaction." *Boston Globe,* 3 April, 27.

Chase, Carl T. 1929. "A Test for Polarization in a Beam of Electrons by Scattering." *Physical Review* 34:1069.

Chikovani, G., et al. 1967. "Evidence for a Two-Peak Structure in the A_2 Meson." *Physics Letters* 25B:44.

Cox, R. T., C. G. McIlwraith, and B. Kurrelmeyer. 1928. "Apparent Evidence of Polarization in a Beam of β-rays." *Proceedings of the National Academy of Sciences* 14:544.

de Grazia, Alfred, Ralph E Juergens, and Livio C. Stecchini, eds. 1966. *The Velikovsky Affair*. New Hyde Park, N.Y.: University Books.

Dye, Lee. 1992. "Scientist Killed, 3 Hurt in Explosion at Research Facility." *Los Angeles Times,* 3 January, 3.

Fleishmann, Martin, Stanley Pons, and M. Hawkins. 1989. "Electrochemically Induced Nuclear Fusion of Deuterium." *Journal of Electroanalytical Chemistry* 261:301. [Erratum, *Journal of Electroanalytical Chemistry* 263:187]

Goldsmith, Donald, ed. 1977. *Scientists Confront Velikovsky*. Ithaca, N.Y.: Cornell University Press.

Jones, S., et al. 1989. "Observations of Cold Fusion in Condensed Matter." *Nature* 338:737.

Jones, Steven. 1992. "Kamiokande Experiments." sci.physics.fusion electronic newsgroup, 3 December.

Kuhn, Thomas S. 1970. *The Structure of Scientific Revolutions*. 2nd ed., enlarged. Chicago: University of Chicago Press.

Langmuir, Irvin. 1989. "Pathological Science." *Physics Today*, October, 36. [From a recording of a talk, presented on December 18, 1953, edited by Robert Hall]

Lee, T. D., and C. N. Yang. 1956. "Question of Parity Conservation in Weak Interactions." *Physical Review* 104:254.

Morrison, Philip. 1957. "The Overthrow of Parity." *Scientific American*, April, 45.

Pool, Robert. 1989. "Fusion Breakthrough?" *Science* 243:1661.

Rafelski, J., and S. E. Jones. 1987. "Cold Nuclear Fusion: The Electronlike Particles Called Muons Can Catalyze Nuclear Fusion Reactions." *Scientific American*, July, 84.

Rhine, Joseph B. 1934/1964. *Extra-Sensory Perception*. With a new introduction. Boston: Bruce Humphries.

Sagan, Carl. 1977. "An Analysis of Worlds in Collision." In *Scientists Confront Velikovsky*, ed. Donald Goldsmith. Ithaca, N.Y.: Cornell University Press.

Velikovsky, Immanuel. 1950. *Worlds in Collision*. New York: Macmillan.

Wu, C.-S., et al. 1957. "Experimental Test of Parity Conservation in Beta Decay." *Physical Review* 105:1413.

Yonezawa, Y. 1993. "Computer Simulation Methods in the Study of Noncrystalline Materials." *Science* 260:635.

Ziman, John M. 1968. *Public Knowledge*. Cambridge: Cambridge University Press.

Chapter 9

Broad, William J. 1990. "Hunt for Aliens in Space: The Next Generation." *New York Times*, 7 February, C1.

Gardner, Martin. 1990. "The Great Urantia Mystery." *Skeptical Inquirer* 14:124.

Gray, Thomas. 1987. "Educational Experience and Belief in Paranormal Phenomena." In *Cult Archaeology and Creationism*, ed. Francis B. Harrold and Raymond A. Eve. Iowa City: University of Iowa Press.

Halliday, David, and Robert Resnick. 1988. *Fundamentals of Physics*. 3rd ed. New York: Wiley.

Kaufmann, William J., III. 1991. *Universe*. 3rd ed. New York: Freeman.

Kerr, Richard A. 1992. "SETI Faces Uncertainty on Earth and in the Stars." *Science* 258:27.

McDonough, Thomas R. 1987. *The Search for Extraterrestrial Intelligence*. New York: Wiley.

Ritchey, H. W. 1977. "Propellant." In *Encyclopedia of Science and Technology*. New York: McGraw-Hill.

Sagan, Carl, ed. 1973. *Communication with Extraterrestrial Intelligence*. Cambridge, Mass.: MIT Press.

Shapley, Harlow. 1963. *The View from a Distant Star*. New York: Basic Books.

Simmons, J. F. L., and C. R. McInnes. 1993. "Was Marx Right? or How Efficient Are Laser Driven Interstellar Spacecraft?" *American Journal of Physics* 61:205.

Tarter, Donald E. 1993. "Treading on the Edge: Practicing Safe Science with SETI." *Skeptical Inquirer* 17:288.

White, John H. 1972. *Early American Locomotives*. New York: Dover.

Wilford, John Noble. 1993. "2 Voyagers Glimpse the Solar System's Edge." *New York Times*, 27 May, 1.

Chapter 10

Aldridge, Bill G. 1992. "Project on Scope, Sequence, and Coordination: A New Synthesis for Improving Science Education." *Journal of Science Education and Technology* 1:13.

American Association for the Advancement of Science. 1989. *Science for All Americans*. Washington, D.C.: American Association for the Advancement of Science.

Arons, Arnold B., and Robert Karplus. 1976. "Implications of Accumulating Data on Levels of Intellectual Development." *American Journal of Physics* 44:396.

Beardsley, Tim. 1992. "Teaching Real Science." *Scientific American,* October, 98.

Bettleheim, Bruno. 1975. *The Uses of Enchantment*. New York: Random House.

Carle, Mark A., Mickey Sarquis, and Louise Mary Nolan. 1991. *Physical Science, the Challenge of Discovery*. Lexington, Mass.: Heath.

Cromer, Alan. 1989. "Early College Admission for Bright Students." *Education Week,* 17 May, 36.

Cromer, Alan, and Christos Zahopoulos. 1993. *Soucebook of Demonstrations, Activities, and Experiments*. Boston: Northeastern University.

Fund for the Advancement of Education. 1957. *They Went to College Early*. New York: Fund for the Advancement of Education.

Hurd, Dean, Myrna Silver, Angela Bornn Bacher, and Charles William McLaughlin. 1991. *Physical Science*. Englewood Cliffs, N.J.: Prentice-Hall.

Krauskopf, Konrad B., and Arthur Beiser. 1991. *The Physical Universe*. New York: McGraw-Hill.

Lawson, Anton E., and John W. Renner. 1974. "A Quantitative Analysis of Responses to Piagetian Tasks and Its Implications for Curriculum." *Science Education* 58:545.

Leventman, Paula. 1992. Project SEED: Evaluation Report. Northeastern University, Boston.

Morrison, Philip, and Phylis Morrison. 1989. "But TV Is Not Enough." *AAAS Observer,* 3 November, 5.

National Research Council. 1993. *National Science Education Standards: An Enhanced Sampler.* Washington, D.C.: National Research Council.

National Science Teachers Association. 1992. *Scope, Sequence, and Coordination of Secondary School Science.* Vol. 1: *The Core Content.* Washington, D.C.: National Science Teachers Association.

Padilla, Michael J. 1983. "Formal Operations and Middle/Junior High School Education," In *Science and the Early Adolescent,* ed. Michael J. Padilla. Washington, D.C.: National Science Teachers Association.

Piaget, Jean. 1929/1960. *The Child's Conception of the World.* London: Routledge & Kegan Paul.

Storey, Richard D., and Jack Carter. 1992. "Why the Scientific Method?" *Science Teacher,* December, 18.

Swartz, Clifford. 1993. "Old Educators' Tales." *Physics Teacher* 31:268.

Yager, Robert E. 1991. "The Constructivist Learning Model." *Science Teacher,* September, 52.

Zeitler, William R., and James P. Barufaldi. 1988. *Elementary School Science.* New York: Longman.

Ziman, John M. 1968. *Public Knowledge.* Cambridge: Cambridge University Press.

Appendix A

Brennand, W. 1896. *Hindu Astronomy.* London: Chas. Straker & Sons.

Appendix B

Carle, Mark A., Mickey Sarquis, and Louise Mary Nolan. 1991. *Physical Science, the Challenge of Discovery.* Lexington, Mass.: Heath.

Hestenes, David, Malcolm Wells, and Gregg Swackhamer. 1992. "Force Concept Inventory." *Physics Teacher* 30:141.

Krauskopf, Konrad B., and Arthur Beiser. 1991. *The Physical Universe.* New York: McGraw-Hill.

Index